Preface

Citizen Education is not a choice; it is a fundamental right. Yet, there is no conscious effort to create a fertile ground for knowledge and wisdom in pursuit of happiness for humanity. It lends to individualism that hinders collective wisdom.

Technology has been at the forefront of innovation for centuries. Artificial Intelligence or AI has dominated much of media and folklore lately. However, we need a reliable and trustworthy system to advance the knowledge of the vast majority of people. One loses the platform soon after tertiary education. As Moore's Law suggests, the computing power doubles every two years, thus requiring the acquisition of new knowledge as frequently.

AI is gaining momentum. It is in our lives everywhere. To the amazement of most, it is infiltrating every facet of livelihood. Yet, the collective wisdom is feeble. The creators are spending trillions of dollars and employing the best minds to outsmart us, the citizens. The sole purpose seems to be wealth creation for themselves at our cost, the winning formula in so-called capitalism.

The time is ripe for picking up our struggle. We have to find this battle in our minds – not on the streets and battlegrounds. Let's educate ourselves; that's our choice.

This book provides practical and real-life examples of AI today. It is suitable for all ages. Neither technical jargon nor AI theory is contained herein. It covers AI in the most relevant eight areas of our lives: wealth, safety, jobs, health, faces, fake, beauty and cars.

Become an informed participant in AI discussions at barbeques, parties, office meetings and other everyday events.

Pick it up, read and gain insights into the everyday impact of AI on many endeavours.

This is the only way we can rise with the creators – those who are implementing AI in the world, not just the business world, to make more fortunes for themselves. Together we stand; divided, we fall.

Once educated is continuously educated. Let's begin the journey towards ultimate Citizen Education, a society where everyone has the opportunity to learn anything at will.

– Prabash Galagedara

Author and experienced Finance, Data and Analytics, and Technology Executive for over 20 years, with Citizen Education at the heart of everything I do.

Embrace

In pursuit of happiness through Artificial Intelligence

Prabash Galagedara

First published by Busybird Publishing 2022

ISBN:
Paperback: 978-1-922691-63-7
Ebook: 978-1-922691-64-4

Cover image: Kev Howlett, Busybird Publishing

Cover design: Busybird Publishing

Layout and typesetting: Busybird Publishing

Busybird Publishing
2/118 Para Road
Montmorency, Victoria
Australia 3094
www.busybird.com.au

Amma and Thaththa (Mum and Dad),
your eternal love, countless sacrifices and
endless support for making an average kid
believe anything is possible!

Contents

Preface   1

Tomorrow is here   3

Repetition is the mother of learning   11

Trajectory   17

Let's understand AI   27

Lesson 1:
Wealth matters   33

Lesson 2:
Wealth protection is no less important   51

Lesson 3:
Your next job   71

Lesson 4:
Health is wealth   89

Lesson 5:
Who owns my face?   107

Lesson 6:
Fakes are getting worse   131

Lesson 7:
In the eye of the beholder   147

Lesson 8:
Our vehicles   167

Will AI take over the World?   187

Notes   197

Tomorrow is here

What a magical place! No miracle, it is called the World's End. There is no sight of the bottom of the cliff. As they envisaged, it is a natural wonder. It was a sunny frosty morning. They were so delighted to make the expedition to get away from their busy schedules and the science lab.

Josh and James were working on a revolutionary project called "Project Oracle". If things go according to plan, they will create the world's first robot with a human brain. Oracle would behave like a human with real emotions. This would pave the way for so many new scientific advancements in the years to come. It would be a dream come true for friends from childhood, Josh and James.

Josh and James were on a deadline to finish the engineering work before the Easter break. If they managed to complete all mechanical and engineering work before the break, then the only thing they would have to do is find a human brain after their well-earned holiday.

They knew this wouldn't be easy, but both were confident they were closer to achieving their objective. They had laid the foundation for most of the exploratory work through one of the leading universities in the country, their alma mater which is at the forefront of scientific discovery. James was even more excited as his girlfriend, April, was to join them on the overseas trip to a magical place popularly knowns as "World's End". April was an expert in Artificial Intelligence and worked for a company that specialised in natural language processing until she started freelancing.

As they expected, the place was breathtakingly beautiful. The three of them explored the natural fauna and enjoyed the fresh air. April was an avid photographer. She was very keen to capture the natural beauty of the place.

"James, can you take a photo of me with the World's End in the background?" asked April.

James nodded and moved towards April to take the camera. There was a nice breeze and James saw something flying off April. It was her scarf, and he rushed to pick it up. He slipped down the rock and disappeared in a flash. Josh saw what was happening from afar. But it was too late before and he could not intervene to save his friend. April didn't see what happened behind her back. She thought James was tricking her for a minute until she saw everyone around her screaming for help.

Someone said to her, "Your boyfriend fell off the cliff." She closed her eyes and sat on the ground.

When April returned to consciousness, many had surrounded her, park volunteers, fellow visitors and Josh. The only recognisable face among them was Josh. It was a whirlwind journey until rescuers recovered James below the peak. They rushed him to the hospital.

April could recall only a few flashbacks in the next few days.

Most had escaped her rattled mind. Josh had made a crucial decision that would keep James's hopes alive. He had to preserve his brain for Oracle!

They boarded a chartered flight for the 9-hour journey back home, accompanying James's body and his brain. It was a roller-coaster ride for Josh and April. Josh spent as much time with April to ensure that she recovered from the ordeal in her own time. The project stalled dramatically while Josh took an extended leave, spending time with April to help her recover at her own pace.

April continued to encourage Josh to give life to Oracle as she could see her James once again. She began visiting Josh frequently at his apartment or in the lab, depending on his whereabouts. With all mechanical and engineering work complete and now in possession of a real brain, the only thing that was required to give birth to Oracle was consent from government authorities. They were delighted with the aesthetics of the design.

First, April and Josh continued their friendship and had each other's back. Eventually, the friendship turned into a romantic relationship. Their love blossomed with everyone finally being relieved of mental health concerns.

Nonetheless, they both were devoted to Oracle. Josh was delighted to see his dream come true, and April was enthusiastic to see her James in Oracle. She thought it was the ultimate gift she could offer in James's name. Five feet and ten inches tall, Oracle was in his private chamber as authorities were taking a prolonged period of time to grant the final approval for the first robot with a human brain in the history of mankind.

Finally, Josh received approval from the National Authority of Scientific Creations, and he was so thrilled. It was a dream come true! They could finally see their beloved friend alive again. James's birthday was a week away. So, they decided to wait for a week until they planted the brain in Oracle. This was the magical movement they were dreaming of.

When James's birthday arrived, April brought a birthday cake with flowers to mark the occasion. Everything was set to make history. Josh took the preserved brain out of the refrigerator and took it to the lab for the 3-hour medical procedure. April anxiously waited in the reception area outside the lab.

She was awakened to a jubilant sound from inside the lab. Everyone was chanting, "Oracle, Oracle, Oracle." April opened the door. Oracle was standing on his own feet, puzzled, looking at faces begging for an answer. He saw April and rushed to her and hugged her tightly! April was elated yet stunned. She shook his hand and moved aside to avoid further embarrassment and unexpected responses from Oracle.

Oracle was alive. In fact, it was James in Oracle's body. Scientists didn't expect the brain to retain the memory from James's life. But Oracle was ultimately James anyway. Even though it was a foreign robotic body, the brain, including thoughts and emotions, was that of James. He didn't know what transpired after the trip to the World's End. He sensed as if he had woken up from a deep sleep. However, April had moved on from the ordeal. She knew it was not James but James's brain in Oracle. She didn't have any emotional connection to Oracle.

Oracle continued to live in the lab. The rigorous and extensive experimentation was underway as it was part of the socialisation process to certify him for external exposure. However, April has flashbacks at the sight of Oracle. This was devastating for both April and Josh. This was not anticipated, notwithstanding repeated suggestions by critics of Project Oracle. Oracle was overwhelmed to see April and

Josh together. He believed they had betrayed the friendship.

This struggle continued inside the lab as well. Oracle didn't obey instructions from Josh. Oracle's actions were erratic and irrational, thus posing a danger to other scientists. He had the physical strength of ten humans, causing fear among the lab staff which led to their unwillingness and reluctance to participate in most activities.

On a frosty morning, April came for a catch-up with Josh at the lab. He was at his office, seemingly working on the computer. April slipped into Josh's office right away and Josh embraced her with a kiss upon her arrival. Oracle observed this through the glass doors which separated the two sections. He couldn't bear it any longer. *Enough is enough,* he thought to himself. He was determined to end this forever.

Oracle opened the door and rushed to the window. Josh and April observed what was going on and trailed behind him. Oracle swiftly opened the window and jumped out. The lab was on the 18th floor of the building. When April and Josh reached the window, Oracle was on the ground, motionless. It was later discovered that Oracle's brain encountered severe damage due to the fall and couldn't be revived.

"That was the end of the story," says William. "I read this story about thirty years ago." Whilst it was more like a fictional story at that time, it is more realistic today. Science has improved vastly, and Oracle is becoming a reality soon. "AI coupled with advancements in many areas such as computing will make more and more innovations possible in your generation."

William, Indie, Lizzy, and Luke (WILL) are a family of four living in a suburban area in Melbourne, Australia. The WILL family is at the centre of this book. Lizzy and Luke are teenage children of 17 and 15, respectively. Their father, William, is a school teacher and their mother, Indie, is a property manager in a local real estate agency. Lizzy and Luke partake in sports, music and dancing. They love travelling across the country and have fun camping during the Australian summer. They are a typical middle-class Australian family by any standard.

They embark on a journey to learn about AI in a way that is unique to them. "AI is a winner-takes-all situation," says the father. They don't want to be laggards. Get on the AI bus with us or stay behind. This book shares their journey, so join them in pursuit of happiness.

Repetition is the
mother of learning

Elegantly arranged dinner table with shining silverware. Nicely tucked table cloth with covers. Red candles on the table. The four chairs are in an inviting position. The family is dressed for the occasion. The warm glow of the red candles provides a romantic atmosphere. The hot meal is delicious, yet healthy and plated beautifully. It looked like a masterpiece of Picasso through the mirror on the wall. The reflection makes them identifiable.

There is the sound of water that mixes with the conversation, making it tougher to follow. The water is dripping down from an artificial waterfall, seemingly with a sunset background. But the only lighting in the room is from the candles – candle flames dance around with the flow of air from the air conditioner, rotating from left to right.

It feels magical; sitting down with the family after a buzzing day of work, study and play is heavenly. The four are sitting around the table, participating in a discussion, each learning from one another. It gets loud occasionally, signifying that it is heated yet entertaining and informative. There is a chairperson. There are no distracting phone calls or messages, thus giving everyone a chance to deeply immerse.

This is a Wednesday evening at WILL's family house named "AI Wednesday," their version of chatting and debating on a defined subject weekly. The agenda is mainly AI and its impact on aspects of life. Repetition is the mother of learning. Discussing something repeatedly makes it stick in the memory. Each week is a new topic appropriately completed through proper research.

As a school teacher, William is exposed to AI. Lizzy and Luke also have plenty of exposure to AI at school and in sports. However, they believe classroom learning is more academic in nature, thus making it difficult to translate to the practical world. They want to learn how to apply it to practical scenarios, so they pick a relevant but novel topic for each Wednesday. While agreeing to the format, Indie suggests that they make special dinner table arrangements to provide a unique and vibrant atmosphere. They unanimously agree, locking it down as the format for Wednesday night.

William believes that "repetition is the mother of learning". The more you discuss something, the more you become familiar with it.

The next question: who is in charge of the discussion? Lizzy suggests that they take turns. The person in charge should research one particular area and share it with the rest. Everyone agrees.

My takeaway

As you embark on your AI journey, AI Wednesday is a concept that could be much embraced, perhaps with modifications as per individual requirements. The day and time could be at your convenience. If weekly doesn't align with your schedule, then make it either monthly or fortnightly. More importantly, it should be done. Timing or frequency is irrelevant.

What if I am single? Well, isn't this an opportunity to find a partner for the discussion? It can be a girlfriend or a mate of yours. There might be many blockades to having an "AI night". However, if you are a keen learner, then you must find a way to do it.

Some of you might be thinking: do I have to set up the dining table like WILL family? Well, make it a special night so that you genuinely look forward to it. If it is another ordinary night, then chances are that you will only practice this once or twice as there is no excitement or intrigue or you may not even implement the concept at all.

The final question is: how do I find different topics for the discussion? Well, we have done the hard yakka for you. This book provides all the necessary information to enjoy the AI night, including topics and content, more than what you need. The only thing you have to do is continue to read the rest of the book. You may extend a topic over multiple sessions. It may help if you ask each participant at your AI night to read the relevant chapter of the book and come prepared for the session. Alternatively, you may read sections of the book at the dinner table. The choice is yours.

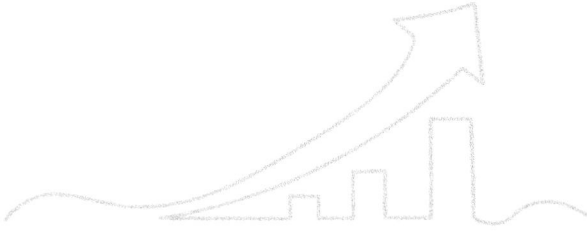

Trajectory

At the age of 11, my father brought home our first television: a 12-inch black and white TV unit sat on a stool in our living area. I still vividly remember the exact date we watched it for the first time. It was on 29 July 1981. We watched the wedding of Charles and Diana live, the "wedding of the century".

My father proudly rested on the couch like David who defeated Goliath. He was a hardworking Australian who only managed to put bread on the table. Everything else was a luxury. The years of savings made it possible to buy small extras at times, and this was one of those moments. As soon as he came back from the TV shop, he said, "You have a knowledge machine at home now, so please use it wisely." TV was at the centre of the family room for many more years until the internet became commonplace.

In disbelief, we sat in front of the small TV and gazed at the wedding ceremony. The world had come to the living room. It was a day that would be in my memory forever; an event in London in the living room of a suburban house in Melbourne was fascinating. It was also a fairy tale to watch – a beautiful working girl from London marrying the first-in-line to the British throne was a magical movement in modern history.

A few years later, my father bought another gadget. My mother's response lacked enthusiasm on this occasion, yet my father's face was beaming with pride and a sense of achievement. She thought it was a show-off and a waste of valuable money on a white elephant. She had a knack for calling out bullshit and was very forthright. She was proven to be correct. We had no knowledge of anyone who had a similar gadget: the telephone. The rotary phone was placed next to the TV for weeks until we received our first phone call on a Thursday afternoon.

There was a ringing sound, something we weren't accustomed to and were not aware of where it was coming from until my father said, "We are getting a phone call."

We all gathered around the living room, but no one dared to answer the call. I lifted the receiver and answered. A soft, gentle voice asked, "Is Indie there?"

"It was the first-ever call I received that happened to be from your father, then-boyfriend obviously," says Indie, sharing her childhood memories with her family.

"Hang on, Mum, how did your parents respond to the phone call? Did they ask who it was?" asks Lizzy, deliberately prying.

"That is a story for another day, darling. But for your information, my parents didn't like it at all, but it was the only way we could connect at the time.

"We have a TV, a telephone and a new phenomenon called the internet all in our pockets now. We call them smartphones. All of this happened just over the last 30 years."

"Mum, did you ever think you would be carrying your TV in your pocket when you first had one?" asks Luke.

"No way," Indie replies spontaneously.

"What will the next 20 years be like? Is there a way we could understand the trajectory of innovation?" asks Lizzy.

Indie pauses for a few seconds. "I think you are ready to listen to the next part of the discussion. Let's talk as we have dinner," she says as she pours red wine to fill her empty wine glass.

Indie explains that over the years, a common set of principles have been developed to measure innovation trajectory and that it is essential to have a basic understanding of what they are for everyday decision making.

"What does that mean, Mum?" asks Luke, confused.

"If you are buying a computer for your school, it is crucial to understand how quickly a new and improved computer will be on the market. You should buy one with a good understanding of performance so that it could be used effectively in the classroom," answers Indie. "There are two important principles we ought to learn. Let me introduce them to you."

"Mum, isn't this AI night? Why aren't we getting into an AI discussion?" asks Lizzy, folding her arms. "It is taking too much time to get into the meaty part of the conversation."

"I can understand your frustration. Remember how you learnt to swim? You wanted to swim to the deep end on your first day, but it took a few years before you could actually do that. We have to learn the basics."

It is essential to understand the fundamentals before delving into a detailed discussion on AI. Machines and computers power AI. Consequently, AI is developed on computers. Thus, it is important to understand the innovation in computing and the cost of such devices. These two principles cover those two aspects. If you don't learn them, it will be challenging to visualise and comprehend the future of AI. These principles will explain the innovation trajectory of computers.

Despite that explanation, Lizzy still seems unconvinced. But Indie proceeds to explain the first principle ...

Computing power doubles every two years

The computing power doubles every two years due to improvements in computer chips. This means our computers can do tasks faster or handle more complex tasks previously not possible every two years. It's known as Moore's Law.

We have witnessed this phenomenon for more than 30 years. The computer that navigated the "moon landing" mission in 1969 was just simply about twice as powerful as a Nintendo gaming console. The first Apple iPhone launched in 2008 was faster than those computers.

"Is that it?" questions William.

"Unquestionably," answers Indie.

Gordon Moore is the author of Moore's Law, an American businessman, engineer and co-founder of Intel Corporation, which pioneered the computer chips industry. His theory revolutionised technology in the 21st century.

If this trend and law continue for the next 30 years, we will witness more powerful computers in our pockets. The increasing power will empower us to either do the same tasks faster or make more complex tasks more manageable. This will also enable more opportunities for AI as faster computing power will make complex algorithms possible.

"My first computer was a desktop that I used mainly for word processing and spreadsheets," says William. "However, I have a computer many times faster today, and not by any chance. I still perform the same word processing and spreadsheet tasks, a routine that has made work productive. But, I can also do more sophisticated and endurance tasks today, such as accessing the internet and running complex programs such as games and robotic programs. That was unfathomable at the time of buying my first computer."

Indie says, "Well, we used to do property advertising in newspapers 20 years ago. The weekly printed magazine was the only medium that was circulated door to door. From memory, we received a copy of the Property Section every Wednesday. Today, all of that is achieved using a smartphone with an in-built camera and a downloadable app. We search property listings online."

This is such a simple concept to remember and apply every day. The power of computers increases every two years, and we can do either the same task faster or much more complex functions that are not possible today.

The WILL family is now ready to learn the second principle. These two are intertwined.

With doubling of production, the cost falls by 10%–15%

This is a law that predicts the cost of producing gadgets such as computers and phones. It states that costs will fall by a constant percentage for every cumulative doubling of units produced. The rule of thumb is 10%–15%, thus making it a good predictor of the future cost of technologies. The theory is known as Wright's Law.

Theodore Paul Wright, also known as T. P. Wright, was an American aeronautical engineer and educator. While studying aeroplane manufacturing, Wright determined that the labour requirement was reduced by 10–15% for every doubling of aeroplane production, which was the foundation for his seminal work later.

Australia's first mobile phone system began in Melbourne in August 1987. The machine was approximately $5,000 and weighed 14 kilograms. It could barely do any sophisticated activities. We have more sophisticated smartphones with a price tag of one-fifth of that today.

"I remember vividly that our school principal used to have one in the early 1990s. It was the size of a brick. Truth be told, it was called the 'brick phone', it was as heavy and shaped like a brick. It was placed on his table like an ornament. The early adopters didn't receive many calls during the first year despite the exorbitant cost of ownership," William chimes in.

"I hope this continues over the next two decades," Lizzy and Luke simultaneously say. They seem so excited. The futurist Sir Arthur C. Clark famously said, "Sufficiently advanced technology is indistinguishable from magic."

"Guys, we now recognise Moore's Law which talks about computing power and Wright's Law which covers the cost of production," says Izzie. "These two are fundamental concepts to understand the future direction of technology. To put this in layman's terms, we will have faster machines at a lower cost in the future. The phenomena will make AI much more accessible and advanced in the future."

"Well done, Indie," says William. "You did a brilliant job learning these two concepts and translating for us with simple examples."

"One more critical point, guys," Indie adds enthusiastically.

Optimism to pessimism

Roy Amara famously said, "We tend to overestimate the impact of technology in the short run but underestimate it in the long run." This is known as Amara's Law.

Amara's Law implies that between the early disappointment and the later underestimate, there must be a moment when we get it about right; pundits estimate it is 15 years down the line. Roy Amara was an American scientist and a futurist who famously coined the above observation which became law later.

This is a very important law that must be read in conjunction with Moore's Law and Wright's Law. All three of them go hand in hand in understanding the future of technologies.

As there is now a good understanding of grounding principles, let's understand the definition of AI next.

Let's understand AI

Before we explore how AI impacts different aspects of our lives, we must first understand what AI is.

In simple terms, AI is a collection of different technologies and techniques working together to enable machines to sense, comprehend, learn and act rationally.

For a machine to "act rationally" means having algorithms at the same level of intelligence as humans or perhaps, in some cases, exceed human intelligence. More importantly, machines need to operate rationally in a given situation. Humans make intelligent decisions based on instinct and experience, which are acquired over time. Values have an impact on decision making notwithstanding previous experiences.

The other part of the definition is to "sense, comprehend, learn and act". These are activities machines perform. Learning and acting are probably the most relevant. Machines must learn and act based on that knowledge to "act rationally".

How do machines learn? Well, this is where data comes in. Data is used to teach machines. Invariably, it is foundational for AI. Machines need abundant data. The machine is given data through which they can begin to learn.

The data has to be robust and accurate so the old adage "garbage in, garbage out" is relevant. If the training data is not robust, then the AI is unlikely to be rational or accurate. Data should also be representative. For example, if the data doesn't include a section of the population, then the AI is likely to make biased decisions.

Machines should act rationally. In its simplest form, AI is a field that combines computer science and robust datasets to enable problem-solving or decision-making. In solving problems or making decisions, machines should act rationally. If machines can't act rationally, it becomes a poor or biased AI model or system. Rationality is defined as the best and most optimal action for a particular scenario.

The word "algorithm" is touted in academia to describe the process of AI. In its simplest form, an algorithm is a sequence of well-defined instructions solving a specific problem or challenge. It has a "start" and an "end" with input derived from structured and unstructured data. The core of the AI algorithm is the logic that improves or learns over time. There are many AI techniques embedded within the algorithm. The learning process encompasses practices such as "supervised" and "unsupervised".

For example, a real estate firm has to predict house prices for clients. The input data includes recent sales with attributes such as land size, number of rooms, location and sales price that determine the houses' estimated market value. As the input data includes the prices of recently sold homes, it is classified as a supervised learning algorithm. The users or developers identify both input and output attributes.

Contrary, an algorithm that identifies attributes that drive a particular behaviour, such as a purchasing behaviour based on groups, clusters or segments, is deemed unsupervised learning. For example, a company identifies types of customers that buy specific products based on recent interactions. The key in this instance is that the algorithm identifies key attributes of the AI model, not users or developers.

The above definition is simple, yet it identifies vital characteristics of AI. However, ongoing research and learning will change the description in the future. This book will continue to use the above description to encapsulate the essence of AI.

Today, a lot of hype surrounds AI development, which is expected of any new and emerging technology in the market. There is the question about the ethical use of AI. Like most emerging technologies, AI should also be used to advance humanity.

Regulators across the world are working tirelessly to introduce laws that govern AI. The European Union (EU) is at the forefront of such efforts with the proposed risk-based framework for regulating AI. The proposal defines artificial systems as

software that is developed with [specific] techniques and approaches and can, for a given set of human-defined objectives, generate outputs such as content, predictions, recommendations, or decisions influencing the environments they interact with.

The rise of AI cannot be explained singularly. There are multiple factors contributing to its emergence. The exponential growth of computing power is one apparent reason. But that alone is not responsible for it. Human thinking has played a significant part. We have pushed the boundaries of the limit of possibilities for centuries. It is one such example where our imagination has opened up new frontiers. Education has undeniably fuelled the hunger for more knowledge and advancement. The roles of governments and corporations have a place in history. While corporations have allocated much-needed capital, the governments have responded positively to technological advancements through stimulating policy actions.

All in all, we have consistently had the right ingredients for a fertile environment. AI is infiltrating many fields; if you can name it, then AI has a role in that field. It hasn't spared anyone or any field, not even the most remote part of the world.

However, AI hasn't taken over the world yet!

Lesson 1:

Wealth matters

Anna was a charming and healthy child until she was six years old. She was born to a very humble family in New York. Both parents were blue-collar workers. One fateful day, her parents learnt that she would never see the world again through her eyes. She became blind due to an illness that had caused her retinas to become detached. It changed the lives of everyone around her. Both parents took turns looking after Anna as she needed help to manage her activities due to her disability. She had to learn a lot of things using Braille. The Braille itself was challenging, but she somehow excelled it through sheer determination.

Her father worked as a janitor in one of the buildings on Wall Street. He came home every day after his second job and told her how stockbroking worked. She eagerly waited to hear the stories every day. That was like her story time, and Anna's father made it a point to create exciting tales for her. She visualised an imaginary world that later became a reality. At the age of 15, Anna found her passion in shares. She would listen to trading channels on her TV and calculate stock prices, movements and other statistics in her head. Her passion drove her to study hard and excel at school.

She finished high school and attended the University of New York. She completed her education with a master's degree in history with flying colours.

After completing her education, she wanted to join a firm on Wall Street. She unsuccessfully applied for many internships. At one such interview, a hiring manager told her that she could never work at a Wall Street firm. He said, "There are three things against you: you are a woman, you are blind, and you have no experience." She knew it wouldn't be easy to achieve her dreams but was very confident that she would be able to achieve it if she stayed on course.

She found a job at an advocacy firm that supports visually impaired people. She enjoyed her work as she started learning how to use technological solutions to improve the lives of visually impaired people. It was priceless learning. Nonetheless, she continued her passion for following the financial markets by listening to programs every day after work. One of the associates at her firm she closely worked with knew of her love and ambition. She encouraged Anna to continue to apply for various roles in finance companies on Wall Street. So, Anna continued the application process.

One day she received a phone call from a brokering company. They offered Anna a placement in their trading section. However, the pay was much less than what she was earning at the advocacy firm. She had no hesitation in taking up the position as it was a good entry job into the world of investing. She found the job to be much more enjoyable. But her co-workers didn't know how to communicate work assignments with her. The first 12 months of the new job were challenging.

She persisted until they found a collaborative way to work together.

Soon after, she witnessed the harrowing 9/11 like everyone else in New York. It was horrific. She realised that some of her counterparts in other firms had not returned after that fateful morning. This sombre mood continued for a while in New York and across the world. It took more time for Anna to adjust to the new environment as she had lost many of her colleagues during the time. She never gave up on her dream of becoming a financial advisor and helping people achieve their financial freedom.

Gradually, Anna became a go-to advisor for most of the clients in her firm. Her drawcard was her ability to listen and quickly calculate financial outcomes much faster than computers. She knew she had found a winning formula. Her clients equally admired her capabilities.

With her newfound formula for successful investments and passion for helping people, Anna found herself in the spotlight in the financial world. She became a regular in most of the financial channels and public media.

She advocated a just and equitable investment environment where all investors are treated fairly and equally. This philosophy was very different from Wall Street's mantra at the time. That was when the 2008 Financial Crisis hit the world. It was very catastrophic for everyone. Many people lost most of their wealth overnight, and some of the renowned firms in the financial world went bankrupt. Due to her sensible approach and philosophy of investing, her firm managed exceptionally well during that time. It was exemplary. She became even more popular among retail investors due to her caring and sensible approach to investing.

In 2010, she decided to quit her job and start her own firm. At first, it seemed like a long shot and a daunting task. She didn't have the wallet to fund a start-up, and no one believed that she had the experience to run an investment firm. However, she was determined to make it work for people, retail investors who are primarily hardworking men and women. Through her connections, she assembled a small team. She found a winning formula to make it work without sourcing money from greedy investors who typically dictate terms and conditions to start-up entrepreneurs.

It was a whirlwind journey of two years as they worked on the project in a small office in Manhattan. The team was determined and hardworking. They launched their product two years after the start of the company. It was an investment platform that leveraged AI and technology to make affordable investments available for average investors. It was a very novel concept at the time. Nonetheless, the platform proved to be excellent for an old model that had failed many people, particularly in the recent past.

Anna became a pioneer of automated and AI-driven investing in the world. Her firm offers AI-driven investing in multiple areas, including crypto. Indie concluded that her platform is cheap and reliable and offers low-cost investment options for retail investors.

"That was a great story, Mum. Is she rich?" asks Lizzy.

"She is, and she allocates most of her wealth to charitable activities."

It is a cold but clear Wednesday evening in Melbourne, another exciting evening for the WILL family. As decided previously, the dining table is delicately arranged for an AI Night. The two children made dinner arrangements with Mum since it is their turn today.

After the introductory story, William starts looking for something. He is moving from one area to the next and is opening cupboards to find something, yet seems unable to locate what he is after. Indie asks him a couple of times about giving him a hand. But he looks resolute in finding whatever he is looking for without help. He wants to make it a surprise for the family.

Spenser, a Blue Merle Border Collie, casually follows William, which seems like a desperate attempt to please his beloved owner despite his single-minded focus on finding whatever he is looking for. His unconscious ignorance irritates family members, particularly his wife. The dog had lived with the family for five years and happened to be a rescue dog adopted by the WILL family when he was just two years old. He is obedient and finds comfort near William as his master. He usually gets his dinner after WILL family. However, Wednesday is a special day for him. He gets his dinner well before the family due to AI Night, anticipating lengthy conversations at the dinner table. As such, Indie prepares his meal along with the kids.

Everyone is at the dinner table anxiously waiting for William. He appears to find what he is looking for. He turns up with a box in his hand. It is like an old box.

"What is in your hand?" asks Indie.

William quietly sits down and clears the glassware in front of him. Luke seems puzzled. "Dad, are we covering AI on wealth today?"

William smiles and opens up a box on the table.

It looks like a game board. "Is that what you were looking for?" asks Lizzy with a disappointing tone.

William opens up a Monopoly game box. No one seems interested in playing a board game. In fact, the children have grown out of board games.

"I have Monopoly in front of us, a board game we all played very fondly as a family. It taught you about money and investing at a young age. This tool has survived nearly 100 years, first starting during the Great Depression. It is a real estate trading game that people play for fun and a chance to be a real estate tycoon. We had so much fun playing it as a family. We used to play it mainly on Friday nights back then. Do you remember the lessons you learnt from it?" questions William.

"Cash is king!" Luke whispers to her sister and father.

"That's undeniably right," acknowledges the proud father. "You have to constantly keep cash in hand in an emergency fund for unforeseen things, including opportunities that may arise from time to time. Lack of cash may lead people to sell valuable possessions like properties at a time the price in the market may be disadvantageous for the seller."

"We got it," says Indie with a big smile on her face.

"Being patient is another one. Is that right, dad?" asks Lizzy enthusiastically.

"That's a priceless answer. We have to buy at the right time. You shouldn't buy every single thing that comes your way. That's discipline," says William firmly.

"What about the adage that you shouldn't put all your eggs in one basket?" questions Indie.

"That is so true, and it has survived the test of time. The next important point is to focus on cash flow. The flow of cash builds up a reserve of cash for future opportunities."

William sips from his glass of water. He is happy with how the conversation is panning out so far.

"I wanted to bring us back to the basics of investing or wealth creation. That's why I wanted to put the Monopoly board game before you. I truly think that no matter how you invest, these lessons are so important in our investment journey. Let me reaffirm that we have to follow these basic principles in the age of AI." William starts unpacking the board game while others eagerly wait for the real conversation on AI.

"Let me break this into two parts. I want to cover investing in trading assets such as shares first. Let's focus on that today. After that, I will cover investing in properties. Let's push that to next week, as these two conversations need a fair amount of time. I also want to discuss money management as part of wealth creation. Money management is essentially cash flow management, how you manage your income and expenses directly influences your investment.

"All of these areas have seen an enormous rise in AI. These trends will shape these industries and bring benefits to the consumers for the foreseeable future. Early adopters will gain much more value from these services. However, it is uncharted territory, and we must tread the waters carefully. I also want to point out that this is not financial advice, so you need to speak to a professional advisor before making any decisions. Mum and I also discussed our options with our financial advisor before signing up for the services that we discussed today."

Investing

"We recently started a small portfolio account with a platform that uses automation and AI. One of my colleagues at work first introduced this to me. I was initially a bit sceptical and nervous about it. We had a lengthy discussion about it. We analysed the advantages and disadvantages of investing in something foreign to us. We swear that we will not miscue the next big trend either. We have skipped so many in our time, so we envisaged this was an extremely low-risk endeavour. We decided to invest a small amount first to test the waters. That's where our journey began.

"Due to the fund's performance over time, we are extremely satisfied with our decision and regret not doing much earlier. Nevertheless, it is working well, and we want to share our learnings with both of you today. So far, we are delighted about the progress even if the online platforms are still scaling up.

"We started merely with $2,000. We thought it was a small amount of money to test something with enormous potential. We set our guard rails on what success means in this endeavour. Those trying something new should have a predetermined criterion; without it, there is no measure to assess success. Once we established that it worked as expected, we gradually increased our investment by small proportions regularly. The incubation period was roughly six to eight months, a period where we were on the fence on promised outcomes. Now we have a significant investment that generates a respectable return. However, we still monitor it regularly to ensure it is on its course. We still have our guard rails to measure the investment against benchmarks. Our corrective measures kick in if the actual performance diverges out of range.

"This fund is an investment in Robo-advisor, an automated technology platform managed through software algorithms. Consequently, the fee structure is very low. It provides low entry barriers with an initial investment requirement of $2,000. It has all the right ingredients for good quality, low-cost investment options. So, we jumped in." William firmly tapped on the table with a big grin on his face. Indie acknowledged it with a thumbs up.

"So, AI-driven investment is here already," says Lizzy.

"Yes, it is indeed," William says with assurance.

Both children are impressed with their father's knowledge of emerging technologies and their parents' willingness to take small risks. These types of facilities offer so many advantages. The platform already takes care of the need to monitor your investments regularly for buy-sell decisions, a gift of automation.

"When I first started off investing in shares, I had to call my stockbroker at a particular time during the day. I couldn't get hold of him most of the time, resulting in endless messages via his personnel assistant. I rarely got confirmations on the same day. The trading notes arrived days later via post, which was the only acknowledgement in some cases. Today, everything is instantaneous as I get confirmations via apps and messages."

"What about the low cost? Isn't that a huge advantage?" asks Indie, directing the conversation to cost advantage.

"It is a considerable advantage, indeed," William answers. "The benefits of cost aren't immediately apparent. However, they add up over the long-term due to compounding."

"Can you explain how Robo-advisors work, dad?" asks Luke.

"Robo-advisors make customised or personalised advice by using technology to automate many tasks a human financial advisor or fund manager would have done manually in the past.

"Algorithms can handle many tasks automatically. These tasks include critical decisions like understanding the risk profile and rebalancing a portfolio of assets. Consequently, we can reduce costs and minimise human error. It's entirely digital; you can say goodbye to endless paperwork and visiting an advisor in the city. You don't have to hold your phone for a prolonged period until your advisor picks up your call.

"Investing with a Robo-advisor is purely a matter of going online and answering a questionnaire covering your financial goals, investment time frame and attitude towards risk. The Robo-advisor will provide you with a statement of advice and recommend an investment strategy. If you decide to invest, the Robo-advisor will manage your portfolio.

"The financial advisor precisely performs these duties much less efficiently. There was a considerable volume of manual paperwork mandated by regulations. We made endless mistakes filling out those lengthy forms, to our dismay. Platforms have taken over all of these activities, which have become seamless and more effortless. It is the digital and AI revolution."

"Does that make sense, Luke?" says William. "I hope that you both can comprehend what I am stepping through. It might be challenging to understand the investment process without prior experience. I am saying that most of the activities are taken care of by AI." Everyone nods in acknowledgement.

"Also, human advisors make judgement errors due to emotions that get in the way of making sound decisions consistently. However, Robo-advisors are consistently executing the algorithm with no emotions and feelings embedded in it. Thus, they are consistent, making monotonous activities faster and more accurate.

"Robo-advisors make investing so much easier for people, especially those who are either time-poor or have no experience making their own investment decisions. It is a match made in heaven for people similar to us, busy and inexperienced.

"It is already a gigantic industry and will continue to grow over time due to all the benefits."

"What are the platforms that offer Robo-advice? Are there any examples you may want to share?" asks Lizzy.

"Robo-advice started in the US around 2010 and expanded rapidly across the globe. There are many services around the world. Stockspot was the first of its kind in Australia, which popularised Robo-advice. Since then, many have sprung up. If you search Robo-advisors on finder.com.au, services such as OpenInvest, Six Park, Raiz, InvestSMART, QuietGROWTH, and Spaceship Voyager pop up. There are other platforms available globally. Betterment, M1 Finance, and Wealthfront are among the different platforms featured in global searches.

"These platforms use many AI techniques but one such emerging approach is deep reinforcement learning. It's a learning process in which an agent interacts with its environment through trial and error to reach a defined goal. This mimics how we learn as we don't always get positive reinforcement, we make mistakes and go through a trial-and-error process to achieve our goals.

"Let's go back to our experience. As previously highlighted, it was one of the best decisions in our investment journey. Our portfolio is growing steadily with good returns generated by the Robo-advisor. It has been an incredible journey so far. We will include you children under our portfolio so that you get a head start and good hands-on experience. We are confident that it will be an excellent experience for your education with much enjoyment and satisfaction through the process. It is today's technology, and it will advance exponentially.

"I want to conclude today with a cautionary statement. AI can make mistakes. As we know, AI has made mistakes in the past, and it will do so in the future too. Nonetheless, AI can learn and improve as defined. We haven't yet mastered AI algorithms. This is just the beginning of the AI revolution. Learning mechanisms have many weaknesses, resulting in many errors and mistakes. So, you must regularly monitor these investments and activities to ensure that your wealth is protected. I have heard and read many stories of algorithms following market trends that result in enormous losses for investors. While we learn the positive side of AI, we also have to understand its downside. It cannot simply be a one-sided story of glory.

"The moral of the introductory story, Anna's story, is that AI-driven investment options are available in most investment categories. These categories include shares, currencies and crypto, among other things. Investors have to make sensible decisions on where to invest and how much to invest. It is not set and forgotten like your superannuation or retirement plan. You need to manage your money carefully."

"That's very true," says Indie, nodding in agreement. "We have covered AI-driven investing."

The WILL family is beginning to get a grasp of AI, what it is and how it works and, more importantly, its practical application in wealth management. As they have done, you can start small and test the waters. If it works for you, you can then add more to the initial investment. That should allow you to be a cautious investor. All of these should be done with proper investing advice.

Let's focus on how to save money for investments and what are the new ways we can do it more proactively using technology.

Managing money

"Don't fear the money machine; harness its power," says William. "Mum and I have a money management rule; we spend 60% of our income and save the rest."

That's where you should start. No intelligent AI can assist you if you don't start off with a primary goal and high-level targets for spending and savings. It is undeniably critical to understand that. We call it the 60:40 rule. We only spend 60% of what we earn. No more, no less, that's it.

"We manage our expenses to the budget," says William. "There are two bank accounts. One is for expenses, and the second one is our savings account. If we can manage the spending bucket effectively, then we go for luxuries afterward. If money is left in the expenses account after managing living expenses, we splash that on things we might not necessarily buy otherwise. We call them luxuries. Once you have a high-level target, tools and apps can assist you.

"We started with our banking app. Our primary bank has a great free app. So we use it extensively. There is a cash flow tool in the app. It has spending categories such as groceries, utilities and meals. It provides an excellent summary of total income and expenses. That's how we keep track of everything. This is not AI, but it lays the foundation for AI. We simply call this cash flow management.

"There are AI tools for personal finance management. We use one of them. It connects to our bank accounts, processes our financial data and analyses and provides recommendations."

"I have heard the term 'open banking'. Can you explain what it is?" asks Lizzy.

"Open banking is a regulation that requires your bank to provide your data to third parties for better services," says

William. "For example, you can request your mortgage data released to a comparison site that can provide the best and cheapest mortgage recommendation for you. It is free and doesn't cost any money to use. More importantly, it is well regulated, so your data is protected and can only be accessed by reputable and authorised parties."

"Why do you use other apps?" asks Luke.

"Well, the banking app only provides a historical snapshot. It may change in the future due to innovation. But now we use an excellent third-party app that provides a forecast of our financial situation. It provides a monthly forecast, rolling for 12 months, of our financial position at any given point in time. The app enables us to plan so many things well in advance. We manage a lot of things based on the forecast, such as scheduling school fees to our budgetary requirements. We also organise our entertainment and holiday plans based on the annual forecast," Indie answers.

"In addition," says William, "it guides us on various things such as where we can save money. Sometimes, we receive alerts and notifications of both positive and negative outcomes. For example, if we overspend on one particular category, it regularly alerts us of our recent patterns. The functionality has enormously assisted us over the years.

"We have also automated regular payments. We call this auto-pilot mode. We invariably manage our finance like a company. All of that is possible thanks to these new technologies."

"Well, how do you identify that this particular app is trustworthy?" questions Lizzy.

"That's a great question, Lizzy. We found the open-banking ecosystem to be trustworthy due to regulatory oversight. It is a government-regulated data-sharing process that is reliable.

That was our benchmark for identifying a solution that is reliable and trustworthy. Fingers crossed, this has worked so far for us. I recommend doing your research to verify any app or software before connecting directly with your banking system. You must take the security of your bank account very seriously. In fact, I would like to cover the security of finance and protection of wealth next. Indie has excellent and practical examples she will be able to share to make a point of wealth protection. Protecting wealth is as important as creating wealth. Some may argue that protecting what you have is probably the most important vehicle for creating wealth."

Lesson 2:

Wealth protection is no less important

Everyone is at the dinner table by candlelight. Indie just sat down with a glass of wine. It is after a long day of work in real estate. She wants to relax and enjoy the time with her family.

Indie then tells the following story of what happened to Erica and Noah as related by Erica herself. Erica and Noah are a young couple. They got engaged three months ago and want to settle down in a quiet suburban area as they found their city apartment too restrictive. They worked with industry experts to find a nice place around this area. They had acquired a decent deposit and finance approval to purchase a home. After months of agonisingly searching and inspecting houses, they found a place both of them fell in love.

Initially, both of them were reluctant, like most first-time home buyers. They were anxious about buying a place that they may regret later. They both had planned multiple things; most were massive financial investments such as a house, a car and a big wedding. They just got engaged, and a big wedding is ahead. So, anyone would be obviously reluctant to commit to another significant investment. After days of zig-zagging, they finally signed on the dotted line. It was a big relief!

They paid the initial deposit from well-managed savings they have built up over the years. Erica and Noah completed the paperwork promptly with a 60-day settlement period. Sixty days is a long time, and many things could happen.

The most unimaginable thing then happened to the young couple. They went on a trip with a group of friends after signing the contract. It was a trip they had planned some time ago with some of their best friends to a beachside holiday resort. Erica and Noah love to travel with their friends two to three times every year like an extended family. They visited a Caribbean island, surrounded by blue waters and sandy beaches. Most of the visitors were young travellers who searched for beautiful landscapes and unparalleled experiences. It looked like the magical holiday they were looking for.

They met another group who were similarly minded at the resort one night. They connected well and partied together for the next few days. One evening, they started playing tag in the pub area after a few drinks. The game extended to phones. They would blindfold one person and ask the person to identify his or her phone in a box - no one could identify the phone properly. It was Erica's turn. She couldn't identify her phone but realised that one phone was missing from the box. So, she immediately removed the blindfold and checked. Her phone was missing from the box. One guy from the newly connected group was fiddling with it and he unlocked the phone with her face ID as he sat right in front of Erica. She first thought that it was a prank. After a few minutes of fiddling with the phone, the guy handed over the phone. She felt uncomfortable but didn't feel insecure as they had been hanging out with Erica's group for a few nights previously.

Everything was going well for a week until Erica received a call from Noah. Noah sounded a bit irritated. Erica first thought that he was distressed with something at work. He was working on an important project, and he had been stressed lately. "Did you transfer money from the savings account, babe?" That was his first question.

"Well, how much are you talking about?"

"The whole amount that we had. We had more than $200,000 in the account," answered Noah.

Erica first thought that it was a joke. But, knowing Noah, she thought there was something serious going on here. Noah has a serious personality with a kind heart who wouldn't prank anyone, let alone his fiancée. So, she confirmed that she hadn't transferred any money. First, they thought it was an error. They both rushed to the bank straight away and met a bank manager. The manager confirmed that money had been transferred out of our account by someone. They immediately challenged that and asked the bank to reverse the transactions. The bank manager explained that a fraud may have taken place, and he would lodge a request to review the transactions and take remedial actions. The manager also mentioned that it might take weeks to recover the money if it was a fraud. They were worried that recovery might not occur before the settlement date. They simply didn't have money to buy the house anymore, as the total amount had been extorted from the account.

Erica called the lawyer immediately and requested that the purchase agreement be cancelled. "Well, it is too late to cancel the contract now," said the lawyer. "It is already unconditional. Once the agreement is unconditional, you have to honour it. If you cancel it now, you will lose even the deposit, and the seller may sue you for breaching the contract. So, we have to find a way to fast track the recovery of your money." The lawyer explained patiently to them a few times. They called the Real Estate agent to cancel the agreement, receiving a similar response as their lawyer's.

All the professionals involved in the purchase agreed to support Erica and Noah recover from this ordeal as soon as possible. It was a huge thing to purchase their first home, so recent events were unbearable on top of other significant things in their lives. They drafted a letter for the bank explaining that they were in the process of securing their first home, and these funds were critical for the settlement of the property. The lawyer also provided similar documentation to facilitate the recovery of money from the bank.

The recovery process was not straightforward. Nevertheless, the bank agreed to provide direct assistance during the recovery process to ensure they were well informed of its progress.

Erica and Noah explored how someone managed to access the bank account to transfer the total balance. They believe the second group at the hotel had stolen the bank login credentials of a few friends while playing tag.

They went through hell during that time. The young couple came out just fine at the end. They were not sure until two weeks before the settlement that they could have their own money back to purchase their dream home. Their bank did a great job of making them meet their financial commitment. It would have been such a bad experience for their young lives without that. They settled the house on time and live there now. Indie finished the story with a smile on her face.

"Well, is this about AI, Mum?" asked Luke.

"It is not AI, but this story is foundational to protecting your wealth. Most people think building wealth is critical. Protecting wealth is just as vital," answered Indie. "There are two things that we should focus on when protecting wealth.

The first one is the use of common sense and the second one is the use of technology. They both are equally important. They are two sides of the same coin."

Use of common sense

Technology has changed many things in our lives. How we do business to how we manage our health has transformed industries. We have relied on government agencies for our protection for decades. They are supposed to be the custodians of law and order. When a robbery takes place in our neighbourhood, we probably look to the police. Sometimes, they catch the culprits. The justice system was working well until we moved our lives online. Nevertheless, some would argue that the justice system works against them: particularly minorities or in some areas. There are so many cases of injustice due to a lack of protection from agencies.

The technological change took place quicker than governments could respond with policy and legislative action. If cyber fraud is committed, we are almost helpless about who to contact and seek protection. If our email account is hacked, there is no direct contact with the email provider. We have to submit a form and hope for the best. At best, we will receive a notification within weeks, and upon recovery, all the sensitive information has already been lost to bad actors.

We have to be more conscious about our safety in the digital age. The definition of safety has changed. It is not about physical security; it is about information security. The most important asset is our digital identity. The ability to safely carry out our activities online is critical. It is as important as food and shelter now. We don't have a government protection mechanism. It is in our own hands.

The first line of defence in the age of the digital revolution is the use of common sense. We call this the SNOW approach. SNOW is not just another acronym; it is the only way to guarantee our safety. Its stands for the following four things we must adhere to:

» **S**trong passwords

» **N**ever share personal information with strangers

» **O**ne-time codes

» **W**atch out for scammers

"We call these must-do activities like exercising for survival," says William.

Stronger passwords are such a critical thing nowadays. It is just like your driver's licence or passport. We usually keep our passports secure, possibly in a vault. But some use their first names or birthdays as passwords. People take it for granted. It is probably the simplest mistake anyone could make.

Nowadays, applications force users to create strong passwords. But we should always do so even if the application doesn't force us to. Creating an unbreakable password and keeping it undisclosed to anyone is critical. The most secretive thing in our lives is passwords. We must keep passwords away even from our family members. It is not a choice; it is an absolutely must-do.

Fraudsters know how to trick clueless people. They call as if they are from a law enforcement agency or a representative of a reputed organisation that we have services with. People unwittingly fall victims to these tricks. Fraudsters use nonface-to-face mechanisms such as calls, emails, and messages to prey on people.

Thus, we should never share information on a call, message or link received. There are enough examples of innocent people falling for scammers, and some of our friends and family have gone through these horrendous experiences.

"It is time to say goodbye to stupidity," says Indie.

One-time code is a mechanism some organisations practice to protect customers. It is commonly known as Multi-Factor Authentication. If your service provider doesn't use this technology, then it is a sign that you are with a service provider who doesn't take your security seriously. So you have a few choices.

Find a service provider that provides better online security. You may wonder why this is so important. If a fraudster is trying to change your bank account password, the system will send a one-time code to your phone for the change request. If you haven't interacted with your bank at that time, it is a red alert to you that someone is fiddling with your account without your consent. You should then act immediately either by contacting the bank or changing the password.

Last but not least, watching out for scammers is a better way to manage unforeseen cyber or fraudulent attacks. People should be conscious of their digital security. Scammers find new ways to trick customers. That's their game. So, we have to be alert and educate people around us to make society more accustomed to the digital world.

"We have a habit of discussing the type of scams at school," says Luke. "There are so many scams on social media. Some fraudsters use fake accounts, and these fake accounts are operated from the world's most remote corners. If you fall victim to these, there is no chance of recovery. These hackers share your information on the dark web for money. The identity of people is a big business for scammers. We have to think about not just traditional channels like calls and emails in this instance. Social media is another area we have to protect ourselves from."

"Hang on, is it just social media only? What about gaming platforms?" Lizzy adds.

Gaming is spreading like wildfire. Children as young as eight years to adults as old as eighty are avid users of gaming platforms. The emotional and sensory maturity of users vary considerably. We have to be watchful on all platforms regardless of where we are. No place is safe. In other words, we have to protect all our digital journeys vehemently.

Use of technology

Common sense will either prevent or minimise our own mistakes. However, it is not merely enough to protect yourself in the digital world. The second part is the use of technology critical for protecting you. Before reviewing the technology, let's understand what needs to happen to protect oneself. Four things are foundational. They are must-do in the age of digital technology.

» Your creditworthiness

» Your devices

» Your passwords

» Your children

Credit worthiness is the extent to which a person is considered suitable to receive financial credit, often based on their ability to repay on time. It can be negatively impacted through activities carried out by others using your credentials; or misusing someone else's credentials. This is often a challenge that many face in the wake of increased cyber-attacks.

People often find, to their dismay, that they owe thousands or millions to organisations in their names due to fraudulent activities committed on their behalf. As in the case of Erica and Noah, the discovery happens when people undergo a significant financial transaction like the purchase of a property. Unfortunately, it is not a good time to realise that credit worthiness has been compromised or money has been extorted.

We own so many devices nowadays. Computers, tablets, phones, gaming consoles and many more digital tools are entry points to access our digital identity. They are similar to the front door of the house. We don't leave our front door open, so we should manage all our devices with a similar mindset.

Passwords are unique codes to protect user access to the web. The more places you enter, the more unique codes you have. In the previous section, we discussed the importance of creating unique and unbreakable passwords. We need to keep a record of these passwords somewhere. It is not advisable to write them on paper or in your diary as people often do. We need a secure and safe password vault for ourselves.

The FIDO Alliance is an industry alliance. Its mission: to develop authentication standards that help reduce the world's over-reliance on passwords. The FIDO Alliance is working together to change the nature of digital authentication or verification which are more secure than passwords. The good news is that there is innovation that will change the landscape of passwords very soon.

Even though the global adoption will take years, we are about to enter an era of passwordless online access. Technically, the FIDO standards will use public-key cryptography to provide stronger verification. When a person registers online, it will create a private and public key. The private key will remain with the person's device. The private key can only be unlocked using voice, fingerprint, face ID, passcodes – biometrics or possessions unique to the person. These technological advancements make the online experience more secure. Nonetheless, common sense will be critical for the success of technological improvements.

Last but not least, protecting children from various digital threats is critical for the health and wellbeing of our future generations. Stories of cyber-bullying of children and the loss of so many lives so early are common. Protection of the most vulnerable in society is not a choice but our responsibility. This is the responsibility of all citizens, not just their parents and governments.

One wonders what the options available in defending our fundamental rights, our digital security, are. We don't have much choice in this area. There is no cyber police and no government-provided security on the web. We, as consumers, have very limited choices. Most large organisations focus on areas where there is commercial value. For example, there are so many security providers for corporations as they can charge a fee from companies. Many individuals can't afford to pay a cyber protection fee.

However, it is essential to allocate some money for cyber protection or digital protection from the monthly budget. "As you may recall, we allocate 60% of our income for living expenses. We should allocate a small amount out of this budget to digital protection," insists Indie.

With a commitment to digital protection and allocation of funds from the budget, we are ready to explore options in the marketplace. As previously emphasised, we have minimal choices. One of the main options is formerly known as anti-virus software. They have evolved to cover many other digital threats such as identity theft, password management, and dark web monitoring, among other things. The most known providers include Norton and McAfee. The investment in this is very small compared to the potential benefits it may provide. These application providers use AI extensively to identify, manage and protect consumers.

One of the popular applications, McAfee, highlights the following on its blog website.[1]

The volume of threats and information that must be processed is greater than humans alone can manage. We need the speed of machines to process, adapt, and scale. Machine Learning (ML), a branch of AI, is commonly applied to the detection process. ML focuses on the use of data and algorithms to imitate the way that humans learn, gradually improving its accuracy. But we need humans as well to outmatch the wits and ingenuity of the human attackers on the other side of that code. We need teams of humans and machines, learning and informing each other – and working as one.

The organisations are using innovation to combat threats from sophisticated and organised groups whose job is to target organisations and individuals alike to profit from vulnerability.

---

1 - V. Varadaraj, 2 June 2021, "The What, Why, and How of AI and Threat Detection", blog article, viewed May 2022, https://www.mcafee.com/blogs/internet-security/the-what-why-and-how-of-ai-and-threat-detection/

It also highlights the top 3 ways AI enhances cybersecurity to better protect online users.

1. Detecting threats

AI has enabled the anti-virus software providers to proactively identify and manage threats to their customers. Early detection of threats and prevention is key for a sustainable digital world. The new way of AI-backed methods is uplifting the traditional human-led processes. Functioning exclusively from either of these two methods will not result in an adequate level of protection. However, combining them results in a greater probability of detecting more threats with higher precision. Each method will ultimately play on the other's strengths for a maximum level of protection.

2. Vulnerability management

AI enables threat detection software to think like a hacker, a phenomenal use of technology to protect people. It can help software identify vulnerabilities that cyber-criminals would typically exploit and flag them to the user. The new techniques also enable threat detection software to better pinpoint weaknesses in user devices before a threat has even occurred, which is a fundamental shift from the traditional methods. AI-backed security advances enhance traditional methods to better predict what a hacker would consider a vulnerability. This is a significant shift in the approach: reactive to proactive detection.

3. Security recommendations

AI can help users understand the risks they face daily and recommend preventative actions. An advanced threat detection software backed by AI can provide a more prescriptive solution to identifying risks and how to handle them. A better explanation results in better consumer education, this feels more genuine than a sales recommendation. As a result, users are more aware of how to mitigate the incident or vulnerability in the future.

"I hope that didn't sound too technical," says Indie.

"It is a very comprehensive review of using technology to protect us on the web. I think we should give Mum a round of applause," says William. The whole family raises their glasses to show their appreciation to Indie.

"Let's finish dinner now. I am so hungry. I think we have to light another candle," says Indie. And that is another AI night at WILL's family home.

•

The WILL family is getting ready for another AI night, which has become a regular activity each Wednesday. William sits down at the table with a box. He doesn't have to look around much this time to find what he wants to share with the family. It looks like another board game. The two children seem uninterested. Indie is feeding the dog. Everyone is waiting until she joins. Everyone is anxiously waiting for their turn. After all, it is an excellent way to learn and focus on family time fruitfully.

"Let's begin another AI night. I am going to start off with a small game. This game is straightforward but informative," says William as he opens the box. "There are two players, and each gets 100 soldiers. You are participating in three wars. You have to allocate your hundred soldiers between the three wars. The party that allocates the greatest number of soldiers to a particular war wins that war and whoever wins two or more wars is the winner for that round. You have to allocate soldiers secretly. Allocation is declared openly once both players finalise the allocation."

"That's an intriguing game. I want to play this game," says Lizzy.

"I will play with her," says Luke.

"Luke wants to be the second player. All good; we have the players now. Let me distribute the soldiers," says William, setting the players up. Everyone seems interested in playing the game. William reiterated the rules of the game as it is a new game to both Luke and Lizzy. Indie seems unaware of the rules and keeps repeating them to familiarise herself with them.

The first round begins. Luke allocates his soldiers secretly. He loads most of his soldiers to the first war with 75, then 20 and 5 to the other two wars. Lizzy is very similar, 60, 20, and 20. Well, the outcome is both win one each. Luke wins the first war, whereas Lizzy wins the third war. The second war is a draw.

The second round kicks off. He is super keen to be the ultimate winner. His allocation this time is 50, 40 and 10, with more soldiers for the first two wars. Lizzy does her part with 60, 40, and 0. She hasn't allocated any soldiers to the third war. However, it is once again a draw. They both are frustrated. Even though they allocate numbers differently, the outcome is the same as round one.

"I don't know why I can't win this game against Luke," says Lizzy, voicing her frustration with the outcome. Lizzy is very competitive and doesn't want to lose anything. She becomes even more competitive when she plays with family members, particularly Luke.

William is ready to add more spice to the game. "Let me provide more information this time, before we start round three. Assume that the three wars are wealth creation, money management and wealth protection. How would you allocate your soldiers now? Let's see your reaction after the new twist to the game."

The third round begins. Luke allocates his soldiers 34, 33 and 33, equally across all three areas. It is Lizzy's turn, and everyone is eagerly waiting for her response. She allocates 33, 33, and 34. Luke and Lizzy then share their decisions. The new rules seem to have changed the perspective of both children. They have allocated soldiers equally. However, each wins one war resulting in a draw again.

"Damn it! It is another draw," cries Lizzy. "Why do we keep getting the same results?"

"Can I ask a question from both of you?" says William. "Why did you allocate soldiers equally in the third round?"

"Wealth creation, money management, and protection are equally important." says both children simultaneously.

"I want to teach you a lesson through this game. You have learned the lesson very quickly. All three areas are so critical in building your long-term wealth. You have to play strongly in all three areas. If you are not trying to win or manage all three areas of wealth creation, money management and wealth protection, then it will adversely impact your wealth. You can't create wealth without properly working money in your favour.

"Remember the 60:40 rule – sixty percent of your money on expenses and the other forty percent on wealth creation. Wealth protection is as equally important. Therefore, managing all three areas is undeniably critical. No protection, no wealth," says William.

"There is no better person than yourself to protect your wealth. No matter how much wealth you build, it can be taken away with a click unless appropriately safeguarded. It is like the three points of a triangle. Remember what happened to Erica and Noah," says Indie, recalling the story she shared last week. Both children nod in acknowledgement and understanding. William smiles, the proud parents looking at each other in satisfaction with how AI nights had panned out.

"There are so many opportunities to leverage AI in all three areas," William continues. "As outlined above, we have discussed some of them in previous AI nights. AI will evolve with time. We should continue to focus on using AI to obtain more wealth. AI will be more common in your time. Keep your options open and use them wisely and securely. If you start with what we have discussed, it will provide a good foundation. Over time, with technological advancements in all areas, there will be more opportunities to leverage AI in wealth management for the benefit of the ordinary person."

"Mum, Dad, this is fascinating. Thank you so much for introducing these topics to us," says Lizzy. "We are learning so much every week. We should continue until we put these new learnings to practice. We are grateful for all that you have done for us."

Finally, governments and regulators around the world are working on strengthening the web by introducing cyber security legislation. This is essential as the digital economy offers significant opportunities for people around the world. It has already opened up borderless trade: international trade without boundaries. An item purchased online would be manufactured in Asia and sold in America to a buyer in Australia with distribution centres located in Europe. All of these take place within seconds.

Closer to home in Australia, the Government opened a consultation on options for regulatory reforms and voluntary incentives to strengthen the cyber security of Australia's digital economy. There is also Security Legislation Amendment (Critical Infrastructure) Act 2021 which creates a framework for managing risks relating to critical infrastructure. All different legislative actions create a framework for managing cyber security.

There are similar actions across most if not all OECD countries. As the digital economy matures, lawmakers across both sides of the aisle are rushing to introduce tougher and more stringent laws to protect the citizen. The US Senate voting unanimously to pass Cybersecurity Legislation Requiring 72 Hour Cyber Incident Notification is one such example.

Law sometimes sleeps but it never dies!

Lesson 3:

Your next job

Financial freedom is broadly defined as "having enough income or wealth sufficient to pay one's living expenses for the rest of one's life without being employed or dependent on welfare". We, as humans, dream of achieving financial freedom as early as possible in life.

The foundation for achieving financial independence is creating enough wealth to generate the required income to sustain the living expenses without full-time employment. In order to gain enough wealth, one should earn a good income regularly. Getting into the right job has never been so important. Another AI night at the WILL family home started more like political rhetoric where a candidate for the next top job is pitching for a dream. It is called the dream job – one's ability to find the most rewarding job every single time a new job is secured.

Michael Jordan famously told the following motivational story:

> I've missed more than 9,000 shots in my career. I've lost almost 300 games. I've been trusted to take the game-winning shot and missed twenty-six times. I've failed over and over and over again in my life. And that is why I succeed.

"As you know, I recently joined a new school. I love my job. It is a dream come true. Indeed, it proved that my best is yet to come. But it wasn't a straightforward exercise. I dreamt of finding a job in a leading private school for nearly two years. It was not an easy task; it frustrated me immensely. Mum knows how hard it was on myself," says William as he looked at Indie with a smile.

"At times, I felt like quitting even before it started. I believed my destiny was to be a PE teacher at a public school. But I persisted. Indie encouraged me to keep going. Jordan's quote motivated me to keep trying. I had to learn many new skills. My endeavour led to discovering AI in the job recruitment process. It was the road to finding AI in the practical world.

"I was a PE teacher for more than ten years in a public school. I love Physical Education. There were so many aspects of my job that I relished immensely. Working with children to improve motor skills, doing physical activities myself and participating in school sports carnivals, among other things, are activities I love the most. It was my childhood dream. I worked so hard to achieve it. Once I got the appointment as a PE teacher, I thought that was it. I finally thought I had reached my dream. It was a matter of living the dream every single day. I certainly enjoyed working with children. The most rewarding part of the job was seeing hardworking children achieve beyond their potential.

"Every year, I found a handful of children who managed to exceed everyone's expectations by simply working hard. One particular year, I worked with a child with no limbs. He had been born without limbs. He was so passionate about swimming. We worked so hard together to teach him to swim. He managed to do that after years of training. That was a magical moment in all of our lives.

"I was in a bubble for more than ten years. It was my dream but deep down, I questioned myself. It was a question about my potential. Is this the limit to my potential? I didn't have an answer. I knew I could do more if I genuinely pushed myself. It didn't occur to me that I could climb to another summit in a different area.

"I began the journey with my old man. He has been a mentor to me all my life. I caught up with him over a meal. 'Dad, I have been doing my job for more than ten years. As you know, this was my dream job. I feel like I am living the dream every single day. I probably have another 20 years left in my career. I am not questioning myself about doing the same thing for another two decades. But I know I can do more. What do you think about it?' He was in deep thought for a while. 'William,' he said, 'since you are questioning what you do now, I think it is time to think about the next phase. You can dream about something else. Think about what you want to do for the next 20 years.' I knew my dad wouldn't give me the answer I badly needed. But I always trusted him to lead me to find my answer. I walked away from that conversation, exploring in my mind what was next for me.

"I wanted to talk to the principal as he has been a mentor for as long as my career. He knew me from my first job and guided me through ups and downs for a long time. I knew he would provide sound counsel on this. On one particular Monday, after my first PE session, I went to his office to find a time for a catch-up. He opened the door even before I knocked. He probably must have seen me walking toward the office. 'Sir, I would like to catch up with you on something important to me. Can we find a time after school that works for you?' He suggested we go to a football game that weekend. I was thrilled. We hadn't gone to a game for some time, and it was time for us to go to a game as we used to. We went to the football game that weekend. The game was a thriller. During the break, he turned to me and asked whether anything was bothering me. 'Sir, you know my current job is my dream job. I have been doing this for more than ten years. If I were to continue, I would be for another 20 years. I wonder if that is the right thing to do.' He looked at me and said, 'Well, I think it is time to think about the next phase of your career.'

"He continued. 'You have been doing this job for more than ten years. I think it is time to find the next thing that ticks for you.' He said that in no uncertain terms. At first, I was stunned but loved his passion and enthusiasm.

"It was a watershed movement. I knew teaching was my passion, so I wanted to continue to teach. I like science, mainly sports science. So, I wanted to become a science teacher in a private school. The real journey began at that point. Ultimately, this led to learning AI, not science.

"I had enough experience in teaching science. I worked as a science teacher for years, mostly during afternoon classes for mature kids. With the professional development and classroom hours I had gained, I was ready to apply for new jobs."

There were three parts to securing my next job.

> » Application process

> » Interview process

> » Onboarding process

AI in recruitment is the application of techniques to the talent acquisition process where AI can learn to shortlist candidates and automate manual tasks in the recruitment journey. The technology is used to automate the recruitment workflow, mostly in the case of repetitive and high-volume tasks. AI recruitment software leverages data to generate insights which then is applied to the list of applicants. The process intends to remove bias and improve the efficiency of organisations. However, critics argue that the bias in data compromises the very objective the AI recruitment platforms intend to achieve.

The lack of robust and accurate data is a hindrance to achieving the AI mission of recruitment automation. Furthermore, human biases embedded in historical data could lead to AI learning the same mistakes throughout its learning process. This is a step that software developers build into the process as claimed by many software vendors: removal of biases in recruitment. To date, this is the most argued point against AI-based recruitment which has slowed down adoption across the world.

Nevertheless, there are many advantages of automation, such as time-saving, standardisation and speed of action. AI in recruitment is spreading across industries and geographical borders. It is an unstoppable phenomenon.

Application process

"I had enough experience in applying for jobs," William continues. "However, that experience was from ten years ago. I knew I had to prepare a resume and a cover letter. Sending your resume to the job agencies and meeting them was the norm back then. I was about to get a rude shock as that barely existed in the 2020s. Job recruitment has changed so much over the last ten years.

"It is mainly done via platforms such as LinkedIn. The initial engagement process had changed significantly.

"First, I couldn't find many job advertisements online. So, I started applying for what I could find, thinking the market had shrunk. The response rate was meagre, and I received rejection letters within a couple of days of my application. I continued to apply without much success, adding further frustration to my defeated mind. I couldn't get feedback on my applications, notwithstanding continued efforts to comprehend what was going on.

"Indie asked me to join a forum one day. I was shocked as I hadn't known there were such forums. So, after hesitating for a while, I decided to join a couple of forums. The rejection rate had skyrocketed by then. It was mainly applicants posting questions on the recruitment process and community members providing direction. Since I didn't know what questions to ask, the forum wasn't that useful.

"I contacted a person who seemed like one of the persons providing answers to many questions in the forum. He seemed caring but wasn't available for a few weeks as he was working on a critical project. I waited anxiously until he was ready for a discussion. I continued my job hunting with similar outcomes; straightforward rejections.

"Finally, I had a chance to sit down with George. He seemed very interesting and very knowledgeable. He followed NBA and was an avid follower of the LA Lakers. We hit off very well as I could talk about any sport. He was very fond of the Lakers and the team's emergence as a contender for the championship.

"George told me that the initial filtering of candidates is done by algorithms. AI is heavily used in the recruitment process, which looks for keywords, among many other things. The application process includes a number of steps including AI-backed applications to identify the right candidate for the job. My resume was outdated. The first step was to get a good resume and a cover letter organised. I could borrow a few good samples from George which was a massive help for a novice job seeker. Alternatively, there are resume writing services one may go to for a fee. I decided to try it myself.

"You also need to have a LinkedIn profile. It is your professional social profile. It enables you to position yourself in the market. It's mainly algorithms you have to impress these days to find a good job. If you can't do that, you may find it difficult to secure a good job. Unlike the good old days when your connections are the foundation, AI now automates most hiring decisions.

"Keyword matching is one such step that AI has taken over. However, the role of AI goes far beyond the matching process. There are AI-driven cognitive tests that identify the candidates with the right soft skills: creativity, problem-solving, managing stress, decision making and collaboration. The applications that identify cognitive skills use gamification as a technique. The candidates are presented with games that capture attributes such as speed and accuracy, among other data points that feed the AI algorithm for the rating of candidates. There are automated interviewing platforms that take care of the candidate interviewing process. There is a standard list of questions that get asked from all the candidates. These applications target volume-based recruitment scenarios to speed up the hiring processes.

"With George's insight, I created a LinkedIn profile. Well, creating one is very simple. But getting it done to a professional standard is very critical. There are so many things one needs to work on. The photo, connections, experiences, certifications, all of those things matter. Professional portrait photography is a huge industry. Most photographers who do family portraits have either discontinued that or now focus on digital photography. Mum gave me a hand to get my profile done. She already had one set up for her work, and she knows how to do it magnificently. We organised a professional headshot. The writing of experiences also facilitated the resume writing process. It turned out to be an excellent profile. George continued to provide advice wherever I needed him. We became good friends.

"The most crucial lesson George taught me was to focus on learning how technology works in the recruitment process. He insisted that algorithms are looking for keywords, my responses, speed, facial clues, among other attributes so I have to adjust my approach accordingly. I prepared myself for any AI-driven application process as organisations use a myriad of different tools in the hiring process.

"I thought that was it. Get the resume and profile right, and you are through. But that wasn't the case. I started getting through to the next step in the recruitment process and was so thrilled but realised very quickly that there was so much more to the recruitment process than the profile."

"I was so proud to see your Dad's LinkedIn profile. He had achieved so much professionally, and his profile was a standout." says Indie beaming with excitement.

Interview process

"I encountered another technological barrier after the success of the application process," says William. "It was once again an algorithm or AI. This time, it was a video interview. Even though my first discussion with George provided a good insight into various tools, I didn't have experience in using any of the new tools. The interviewing process had become automated since my previous experience a decade ago.

"The interview process was driven by the humans last time I had an interview. The recruitment agency was taking care of the filtering of candidates. Once the agent was comfortable with the candidate, the hiring manager would take on the shortlisted candidates for the final interview. The candidates would be rated based on the interviews, and the top candidate would be offered the job. If the top candidates weren't available, the next person would get the offer. That was simply the process, and we knew how that system worked. The HR manager or the agency manager would be critical in the recruitment process. They would provide expert advice to the hiring manager. I couldn't believe that AI had taken over that. If someone had said that algorithms would be interviewing me for the next job ten years ago, I would have dismissed that as a joke. No one could imagine the level of automation and dominance of AI in such a short period. I knew I had to catch up with George once again. So without hesitating, I called George and asked for a catch-up.

"We met for lunch in a busy café in the city. George started with basketball as usual. LA Lakers had won the championship that year, and he had gone to watch the final series. He was so happy about his time in the US. George had taken that as an opportunity to learn about emerging technology in Silicon Valley. I felt like he had grown so much as the conversation continued.

"'William, facing interviews is not as simple as recording a video,' said George. 'More preparation is required as the algorithms look for certain things in the process. You should practice interview questions more and be ready for the interview process. The posture, word, and display of certain attributes such as confidence are key for successfully getting to the next phase. There are so many candidates providing short interviews for a job. The system asks the same question from all the candidates. The best video performance is usually selected. So, practice is key to success.' That was short but sharp advice from George and was immensely useful for the interviewing process. I walked away thinking I got this."

"Video-based interviews have many advantages. The speed of recruitment is the main advantage as videos can be recorded at the leisure of candidates and uploaded anytime. It has cut down unnecessary travel and analysis time and has immeasurably increased the number of simultaneous interviews. I can see so much value in the automation and speed of action. It sounds like a great innovation. We conduct property auctions online. Online auctions gained so much popularity during COVID as people couldn't attend in person due to lockdown restrictions," says Indie, speaking from her experience in the real estate industry.

"Another advantage of AI-driven interviewing is reducing hiring-bias and increasing diversity," says William. "There are two sides to this discussion. AI has provided consistency to the process. Everyone is evaluated based on the same process. In contrast, human-driven processes have biases based on the interviewer, previous experiences and other personal preferences. AI also has these weaknesses, but developers consciously add corrective steps to remove known biases.

"AI is also based on historical data, in this case on previous interview data. If the previous data has a bias toward one particular gender, AI will continue that bias."

"One obvious advantage is consistency, the process is common to all candidates. AI has to be created without any of these weaknesses. It is the responsibility of creators to deliver ethical AI," Lizzy adds. "This is probably the most challenging task of our generation. How do we use technology without the same issues of the non-technology era? Dad, what was the tool you had to use for the interviews," asks Lizzy, curious and wanting to do more research.

"There were a few that I used," answers William. "One of the main platforms was HireVue. It was a great, intuitive platform. I am sure all of these platforms will gradually improve over time. The following ten years will be exciting as we will have so many innovations in this area. Recruitment or hiring is one of the first areas AI has transformed. The hiring and recruitment platforms use multiple AI methods to deliver a great hiring experience. Natural Language Processing (NLP) is a branch of AI that enables machines to understand texts and spoken words in much the same way human beings can. NLP draws from many disciplines, including computer science and computational linguistics, in its pursuit to fill the gap between human communication and computer understanding.

"I took George's advice very seriously. I came back home determined to do well in my video interviews. However, I was super keen not to lose authenticity. So my journey continued, and I started getting more positive results from the interviews. Towards the end of the process, I was like a professional who knew what to do. But when I started the process, I was like a fish out of water.

"Finally, I landed my dream job once again. It has been almost two years since I started as a science teacher. I love it! This job is so fulling that I think I will continue to evolve as a teacher for older children."

"Glad you found your dream job once again," says Lizzy. "Was there much AI in the onboarding process?"

"There is so much AI in the onboarding process, particularly training and development. I am going to share my experience next week. How about we finish dinner now?"

Onboarding process

"The job onboarding process was seamless. It included induction, training, system access and many other steps to make the new recruit familiar with the organisation's culture," says William. "It was a very quick but friendly process managed by the principal's office. The school had a good onboarding program. Introduction to the school community was enjoyable. I had so much fun. The school had an excellent training and development platform. At the centre of the platform is AI. It was the best platform I have seen in secondary education. It is fully integrated into the teaching. I felt the impact of AI in three different ways."

William then lists and explains them:

» Personalised training

» Saving time

» Automated learning integrated into the teaching process

"The term 'personalisation' has been used across the business world for some time. I experienced this first hand in my training modules. I am a learner who enjoys visuals like videos and drawings on paper. It can be a repetition process. That's how I learned most of the things during my university days. I listened to the lectures and then took notes in a picture format, making it easy to graph things much faster.

"After a day or so of using the platform, it offered me videos as a learning pathway. I was so surprised that it understood my behaviours so quickly and offered a personalised solution for me. It was so much fun to do training. I don't think a lot of people enjoy training. I had a completely different experience due to the personalisation

"I had a chance to speak to a few teachers about the training platform in the first few weeks. They all said that it was one of the school's best things. I wanted to know about the program. I got permission from the principal to speak to the company that owns the learning platform. He was generous to offer the contact details of the person managing the school's relationship. I didn't hesitate to contact the person. He was very forthcoming in connecting me with the Head of Platform Development.

"It was a midnight call as the person was located in San Francisco. But that didn't matter to me. The guy was fascinating and a visionary. He shared most of what is yet to come for the next two years. I felt like he was someone who could talk to walls for hours.

"AI in learning refers to either a computer program that can simulate human capabilities, such as shape recognition, image recognition, sound recognition, machine translation, man-machine dialogue management including question-answer management, or processes an enormous amount of data to guide decision making. That is what we are working on. The system can learn very quickly what personal preferences are now. We call it personalised learning. You already have that. But the content is somewhat similar to all the users.

"In the future, the system will be able to understand the different needs of individuals, and it will create a training program that meets individual needs. If your school has 100 teachers, it will create 100 training programs to suit individual needs based on a person's knowledge and abilities. AI will make learning enjoyable and much more effective for organisations. It will also have much more sophisticated interfacing like speech recognition, facial recognition and many other methods that even don't exist today.

"Emerging AI methods such as Generative AI will make a significant impact on future of learning. Generative AI is a branch of AI that focuses on generating content like text, images, music and text-to-image generation. In other words, it runs on algorithms that identify the underlying pattern of an input to generate similar plausible content.

"He wanted to share much more with the school community, particularly students. So, I agreed to speak to the principal and organise an event for the school. I felt like we connected well. That was an eye-opener for me. While we are working and sleeping, there are people in the world who work on future technologies to make this world a better place," says William.

"Well, there are factors we must consider in building a future for all to learn and develop. Algorithm bias is one such thing we need to be conscious of. If data is skewed, it can harm the results or outcomes. For instance, if your data favours a particular gender, race or educational status, then your machine-learning algorithm may interpret this as a determining feature or an outcome. To counter this, you should utilise machine-learning systems with anti-bias features. It might include options to weigh opinions differently according to hierarchical levels or creating recommendations by category group. You should also ensure someone checks the data used for AI, identify potential sources of bias, and then scan the results for bias.

"Finally, you will need to educate your analysts about how your AI system works, how to interpret its output and when to dismiss it.

"Another potential challenge is the privacy of data. A lot of this data will have to come from your learners. Some of this data is private and should be treated as such. It means applying all the required privacy measures and adhering to relevant regulations. Furthermore, none of this data should fall into the hands of people who don't need them, whether it is for internal or external use. Therefore, you should seek to take all measures that make your data secure."

"Well, that's a good focus area. It sounds like the future of education will be very different from what we have in schools today," says Luke. "I wonder if these systems will expand into school education. Each and every student will have a personalised learning pathway based on individual preferences. That will be amazing."

"The current system is much better than what we had. Our system was very textbook-based. Currently, your study programs are driven by software applications in schools. However, there is zero to minimal AI in these applications," says William. "However, the future potential is enormous. The next generation will have much more sophistication in learning and development. Your generation should build that for them."

That ended another AI night in the WILL family home. Having a good meal with a conversation by candlelight about the future has become a practice at WILL's family house.

The government's role in AI hiring has been very limited for some time. Many organisations have increasingly adopted AI in talent management and acquisition processes. Responsible AI is a topic that will require the implementation and measurement of success. A growing wave of regulations will soon hit developers and users alike. Over the last year, regulators, including international bodies and OECD governments, have begun to focus on the potential for AI to cause harm. Recently, lawmakers introduced legislation in both chambers of the US Congress requiring organisations to use AI to perform impact assessments of AI-enabled decision processes, among other requirements.

The European Commission, the executive branch of the EU, introduced a proposal to regulate AI, the Artificial Intelligence Act. The landmark act is the first-ever attempt to enact a horizontal regulation on AI. The legal framework focuses on the specific utilisation of AI systems and associated risks. The EU Commission proposes to establish a definition of AI, which is technology-neutral, to classify AI systems with different regulatory requirements based on a risk-based approach. The lower the risk, the lighter the transparency requirements are and vice versa. The proposed legislation is currently at the consultation stage with stakeholders, which will go through a rigorous process before becoming law.

Australia doesn't have specific laws regulating AI, big data, or algorithm decision-making at the time of writing. However, a range of other laws and legal concepts indirectly shape the adoption and implementation of emerging technologies. Relevant laws include privacy and data security, corporate law, financial services regulations, intellectual property law, competition law and anti-discrimination law.

Meanwhile, the New York City Council passed legislation regulating AI tools used in hiring. The statute carries fines for violating the law. The law, due to be in effect in January 2023, calls for audits of tools that automatically screen job candidates. The regulation requires an audit for potential bias. This is the start of a wave of similar regulation in other geographies.

The AI in jobs is about to get very interesting over the years!

Lesson 4:

Health is wealth

Heather was working long hours. Her family had gone on a holiday a few days earlier. She had some urgent work to finalise, so she decided to work an extra one week before joining them on the skiing holiday. The work had been enjoyable, and her team was closer to finalising a new product that was expected to go through extensive testing in 3 months' time. The extra week was crucial as her team was on the verge of finishing most of the work. The last few days had been exceptionally long. She usually started work before 8 am. It was a 30-minute drive to work and finished on most days around 5.30 pm. But this month had been very hectic.

Heather had been working on a start-up company that worked on a vaccine for COVID-19. She was in charge of laboratory work. Her team was instrumental in getting this into the hands of testers to ensure it met regulators' deadlines. She had been working tirelessly to meet the company's internal deadline.

It was a Thursday evening, one more night before she was to join her family on holiday. She kept getting photos every day, being updated with what was happening at the ski resort. The family had to postpone the holiday due to a strict lockdown and restrictions on overseas travel. The weather had been awesome. She glanced at her watch and realised that it was already 7 pm. It was time to go home and have a good rest. Tomorrow was to be the last day before she wrapped up. There was only paperwork left, so Heather was confident she would be able to manage it mostly in the morning. There were a couple of her team members still working. She encouraged them to finish up as soon as possible and go home.

Heather felt a bit tired that day. It was unusual for her. She was usually very energetic, even after 12 hours of work, intense exercise and household duties. Surprisingly, she didn't have the energy to go to the gym. So, she picked up dinner from a fast-food restaurant. While having dinner, she called her family to say that she was excited to join them in two days. As she was getting too tired, she decided to sleep early.

When she woke up, she was surrounded by her family, including her husband and two children. At first, she thought they had surprised her with a trip back in order to accompany her to the ski village. *It can't be,* she said to herself. She noticed an unfamiliar face. It seemed like an unfamiliar place as well. She understood something had gone wrong.

She summoned her strength and asked people around her. "Can you guys tell me what happened?"

The stranger started talking first. He looked like a surgeon. "Well, you are so lucky. You are in a hospital. You are getting better now. In a couple of days, you can go home but can't go on your skiing trip this time. Unfortunately, you have to postpone at least six months. You need to have family time now, so I will come back later. Please don't worry anymore, guys. I am in the office until 6 pm today if you need me. You can talk to me anytime." He made a quick check on her pulse before he left the room.

Heather had fallen asleep quickly previous night. She felt unusually tired. In the middle of the night, she woke up with a bloody nose. She wanted to check, so she got out of the bed and walked to the bathroom. That was all she remembers. She fell over in the bathroom. It must have been a hard fall, face down.

While she was losing consciousness, her husband called a few times as he received a text message. A mum of two with a good job and a nice family – is this how I was going to go? She must have thought that before she fell unconscious. There wasn't anyone to rescue her or take her to a hospital. It was a miserable place to be in for anyone. Life throws so many unexpected events at you. You don't know when or how they come at you. If she knew, she would have stayed at the office that night or perhaps she could have gone on the ski trip with the family.

She wasn't that unlucky. There was one watching her. It wasn't anyone who she thought would be her saviour. It was the most unexpected thing: her watch.

She had a digital watch that monitored her health. She was a big fan of healthy living and physical fitness, which were her priorities. Apart from monitoring exercise and sleep, her watch was barely useful. At least that was what she thought. That wasn't to be the case.

The watch has a feature that detects falls. It alerted the emergency contacts immediately. The first response of fall detection was to confirm if it was, in fact, a fall. There was no response from Heather. Then the watch went into action mode. It rang the emergency services immediately. The paramedics picked up the call and went to assist her. Her husband also called her phone a few times to see what was going on. There was no answer from Heather, so he knew something had happened. He also called the emergency services after. When he managed to get through, paramedics were on the way. He managed to direct them faster as they needed the home's address.

When paramedics arrived, they had to break into the house as there was no response from her. When they managed to find her in the bathroom, she was motionless. They treated her immediately at her home and took her to a nearby hospital.

"That's how it all happened," said her husband. "The watch is the hero, but everyone else reacted very quickly."

It is another AI night at WILL's family house, which are getting more thought-provoking and enjoyable. Some of these stories are practical to help everyone understand a very technical and sophisticated topic like AI.

"That's a lovely story. Why don't we hear stories like these more often? Why can't these manufacturers of digital technologies tell these stories?" says Lizzy.

"That's absolutely correct," says William, giving her a thumbs up. "Let me continue the story. The watch saved her life. If not for the watch, she would have died in the bathroom. Most digital watches have a fall detection feature. At the centre of technology is AI. Healthcare is a brilliant case of using this technology to save lives every single day."

William goes on to explain that fall detection systems use accelerometers, a type of low-power radio wave technology sensor, to monitor the movements of the user. Some fall detection devices use a built-in tri-axial accelerometer with patented algorithms. The fall alert detectors can detect abrupt changes in body movements. The technology can evaluate an individual's body position, physical activity, and the smoothness of acceleration of movements. If the device determines that these variables are within the danger zone and a fall has occurred, it sends an alarm, including vibration that alerts the user. You get the option to call emergency services if you're in trouble. But if the watch detects no response, it automatically calls the emergency services and emergency contacts and plays audio informing that you are in an emergency.

Fall detection was first introduced in watches around 2018. Since then, it has been carried over to newer models. Over time, the companies at the centre of developing technologies have improved the feature to make it more accurate and avoid false detections.

"That's a very clever use of technology and AI. I am sure this has saved so many lives since the first introduction," says Luke, impressed by his father's knowledge and stories.

Health for everyday living

"Health is wealth, as Dad says. It provides a foundation for a good life. A quote on our school notice board says, 'Time and health are two precious assets that we don't recognise and appreciate until they have been depleted.' AI has been at the centre of health for some time. In recent years, wearable has become a tool for good health practices," says Lizzy.

"The watch is more like a radar screen for everyone. From a health point of view, it can provide many alerts on health-related conditions. While so much more can be done using the app, it is probably the only place where we have a lot of information about our health in a single place. I am going to highlight the use of AI on the watch. Luke will cover other futuristic technologies," says Lizzy, setting the agenda for her discussion.

Lizzy goes on to talk about the watch, mentioning that there are many features on a watch now: activity monitoring, body measurements, hearing and heart-related measurements, nutrition; there are many more things one can do with intelligent wearables. It has become so crowded with features that people probably don't even know that these features exist.

The devices collect data and then use algorithms to identify any unusual changes in patterns. The notifications alert the end-user. It's done at a rapid pace due to the speed of processors. This type of automation was probably not possible ten years ago. However, as highlighted by Moore's Law and Wright's Law, this technology has become accessible to everyone.

There are a few essential and critical features available on watches since the launch of the technology. There will be more available in the future as wearable technology is on an upward trajectory.

Heart rate notifications are another essential feature. The watch scans the heart rate constantly; unusually high or low heart rates could be signs of a severe heart condition. The monitor could help people identify situations that may require medical attention. If a patient's heart rate is above 120 beats per minute or below 40 beats per minute, the user will receive a precautionary notification to alert the user. Users can adjust the threshold or turn these notifications on or off as they wish. All heart rate notifications, date, time, and heart rate can be viewed in the health app on the phone.

"Let me share this story about Kelly," says Lizzy.

Kelly was young and was in her mid-twenties. She was very fit and athletic. She never thought she would encounter a medical problem at a young age.

One night after dinner, she was sitting with her husband watching a movie. He alerted her to a notification on the watch that said 120 beats per minute. At first, she thought it was a false notification. She wanted to switch off the alarm. However, he was adamant that she would take note of the message. Throughout the night, her heart rate kept increasing. 'It was so annoying,' Kelly said later on. However, her husband insisted that she visit a doctor the following morning.

The doctor conducted a few tests, but nothing serious could be found. Nevertheless, she was referred to a specialist. She was at first very reluctant to visit another doctor as she didn't believe she was sick. She, in fact, thought that it was a false notification.

She visited the specialist the following day and underwent a thorough check-up. She was diagnosed with a rare condition. If undetected, it would have turned into a life-threatening illness. Luckily, her husband was very vigilant and proactive. She wouldn't have visited a doctor otherwise.

"I think that's a good example of how some AI-driven features save lives. I want to highlight a few more things available in wearable devices," says Lizzy.

Lizzy goes on to explain another feature called irregular rhythm notifications. It checks for signs of irregular rhythms that may suggest atrial fibrillation (a-fib). A-fib is an irregular and often very rapid heart rhythm that can lead to blood clots in the heart. A-fib increases the risk of stroke, heart failure and other heart-related complications. This requires immediate medical attention and treatment.

"That's a great find, Lizzy. Can we check our watches to see if these notifications are on?" says Indie, wanting to put her learning into practice immediately. Everyone quickly checks their watches. Most of the features haven't been activated, so they all quickly activated them.

"The next feature is the most amazing thing I find in watches, the electrocardiogram or ECG app. We used to get appointments with a pathologist to get an ECG. You can do this on your watch nowadays in under a minute," says Lizzy with excitement as she shares the newfound feature.

With the ECG app, people who receive the irregular rhythm notification can capture an ECG and record their symptoms. This real-world data enables users to make more informed and timely decisions on seeking immediate medical attention and care. The app uses the electrical heart sensor built into the watch to record an ECG.

The ECG scan in the watch then provides the sinus rhythm, atrial fibrillation and heart rate, the results available in the health app on the phone. It prompts the user to enter any symptoms such as rapid or pounding heartbeat, dizziness or fatigue. The waveform, results, date, time and symptoms are recorded and can be exported from the health app as a pdf file to share with a clinician. If the patient notes indicate a severe condition, then they are prompted to immediately call emergency services.

"At the centre of all of these are algorithms or AI, which is making so much impact on people's lives in a positive way," says Lizzy optimistically.

"These technologies will continue to evolve. AI will continue to be at the forefront of what is coming out. As seen so far, our watch will be the future doctor - doctor in our pockets. It will track and monitor our health and prompt us to go and see a human doctor if and when required. The future of healthcare is in the hands of the individual."

"That was a fantastic update," says William. Everyone applauds her for her straightforward perspective on what we can do to improve our health today.

"Let's move on to the next area, the future of healthcare. I'll leave it to Luke's capable hands," says Lizzy.

Future health

Luke relates the following story of a middle-aged couple who discovered cancer ten years before they had it.

Jack and Jane are a married couple with no children. They enjoy adventurous activities such as trekking and hiking. They work hard for 3–6 months at a time and take extended breaks to conquer places they loved.

While on an expedition, they met a scientist who worked for a start-up company that focused on futuristic pathology services. While the company's work sounded exciting, Jack and Jane thought it was a gimmick that a group of nerds claimed to have worked on. The expedition went on, and they became good companions over the period. The scientist kept talking about the company's success and wanted to introduce the technology to anyone willing to participate. He offered that as a free service.

At first, it seemed like a straightforward task of participating in a few tests. So, they agreed to take up the offer once they were back home. The scientist booked them up without further questioning for an assessment immediately.

On the day, they arrived at the laboratory on time. They were greeted by their friend at the entrance and accompanied to the testing area. A nurse welcomed them to the lab and completed the sample collection process. They walked away thinking that it was a straightforward process. Their friend reminded them to come for the next session, which was to discuss the outcome of the lab testing.

They did return later, more excited about meeting the friend than sitting down for a discussion about their health for the next ten years. They thought they were very healthy and strong people who didn't need any of this medical hassle at this stage of their life. They both were 40 years old at the time.

Their friend greeted them at the entrance like the last time. This time, he also agreed to sit down with them together with the doctor to go through the findings. It wasn't a big deal for them as they only decided to do this for their friend out of courtesy rather than any medical need.

The doctor sat them down at the computer. The computer was connected to a large display unit so that they could see the small graphs properly. She started explaining the process.

"We took your samples and analysed them in our lab. It provides a current status of conditions. Then, it predicts the change in conditions over the next ten years. We have found this prediction to be very accurate.

"How accurate?" asked Jane.

"More than 90 percent," the doctor answered. Their friend then explained that they collaborated with government agencies and the research sector to develop an algorithm based on historical data for a considerable period. The result was game-changing and exciting.

"There are three tabs here," said the doctor. "The first one is a snapshot of your current health. The second tab highlights the next five years, and the third one is 5 to 10 years. We have arranged it this way to explain this very clearly to our clients.

"Let me open the first tab of your profile, Jane. Your current situation is very healthy. In all respects, you are within the threshold. Congratulations on your current health. Let me go to the next section. Well, it predicts good health but highlights one area for monitoring. Let me come back to it." She opened the third tab. "The system indicates breast cancer for you at the age of 48. You may see some signs much earlier. You have a good chance of managing it very effectively as we have this insight about what may happen in 8 years."

At first, Jane looked annoyed. In spite of everything, she didn't expect the test to find anything at all. Her objective was to help their friend with the start-up. "What can I do?" Jane queried.

"We strongly recommend you a preventative plan now. You have enough time to manage it effectively. However, if you wait, it might be too risky. As a medical professional, I recommend you start a preventative plan sooner rather than later," said the doctor.

Jane was somewhat upset and didn't want to continue the conversation.

"If you have no more questions, let me check your husband's profile." The doctor hopped on to the next patient's profile on the computer. "Let me open the first tab. Well, you are excellent, but your cholesterol is borderline excessive. It is not in the danger zone."

"I kept telling him all the time," said Jane in agreement with the doctor.

"The next five years are looking fantastic, notwithstanding minor issues with the level of cholesterol. There are no signs of any severe medical conditions. Oh, ten years is not so bright. It predicts prostate cancer at the age of 50, which is common in men. We need to be more cautious as your current cholesterol level is also at an elevated level. Let's discuss your preventative plan as well. We don't have to do it today. It is normally frustrating for some people to hear about future medical challenges."

"You may absorb the information today and come back next time for a proper consultation on preventive health measures you may have to undertake." The doctor concluded the session with another booking in 5 weeks for the next steps.

Both Jack and Jane were visibly upset but weren't sure what to do in this situation. The consultation was not about their current health. It was about what may happen in ten years; what they should do now to prevent it. After all, they were sceptical about the reliability of AI-driven pathology. They had a quick chat with their friend and went home without deciding on the next course of action.

They went home and had a good chat. Unsure of the reliability of AI-driven pathology, they decided to skip the next session. However, their friendship with the scientist continued. They continued their adventurous lifestyle.

As time went on, they both experienced cancer in their late-40s. They were swift to get on to treatment plans that prevented the cancer from getting to an advanced stage. They were unhappy and angry that they didn't take the opportunity to do something earlier in their lives to prevent this.

Jack and Jane decided to tell their story to educate future generations about trusting the predictive healthcare system. They became advocates and have featured in many articles and stories highlighting the importance of futuristic and AI-driven healthcare. They still go on adventurous hiking and trekking expeditions.

"Is this a real story?" asks Lizzy.

"It is indeed a real story. I found this on a health forum, and they have become advocates of predictive healthcare. You may search for it yourself," says Luke.

"Lizzy and Luke, thank you so much for your research on the use of AI in healthcare," says Indie. "You both did a brilliant job in simplifying how these technologies help humanity. Indeed, you have put a face to AI in healthcare."

"In a nutshell," says Luke, "AI is helping us live better and longer today, making predictions about our lives in the future. I am genuinely optimistic that these types of healthcare will be available to every person on earth. I don't think it is a privilege. Every government must make healthcare smarter for a healthier and more prosperous life."

"I also read an article from 2017 about the use of AI to prevent blindness. CSIRO, the Australian government agency responsible for scientific research, has developed an eye-scanning technology that enables doctors to test diabetic patients for diabetic retinopathy, a debilitating condition affecting one in three diabetic people that can lead to blindness if untreated. Currently, only specialists can screen for the condition. This advancement is an excellent example of the essential role science plays in finding innovative ways to help people live longer and happier lives."

The CSIRO on its website states the following, which Luke reads.[2]

With this world-first innovation, our scientists are at the forefront of using artificially intelligent technology to save people's eyesight and make healthcare more accessible for all Australians.

The technology's creator and trial co-lead said the innovation could help people with diabetic retinopathy receive treatment faster.

2 - CSIRO, 12 September 2017, "AI technology to help prevent blindness", news release, viewed May 2022, https://www.csiro.au/en/news/news-releases/2017/ai-technology-to-help-prevent-blindness

"Patients at risk of this condition would usually be referred to a specialist for screening, waiting six weeks or more — now it can potentially be done in a single 30-minute visit to a GP," Professor Kanagasingam said.

...

Importantly, as a basis for comparison, the images were also analysed by an ophthalmologist, and the technology was found to be as effective as the specialist in detecting signs of diabetic retinopathy and grading its severity.

"Early detection and intervention for diabetic retinopathy is key, and this new tool is the first step to help GPs prioritise patients for treatment," Professor Kanagasingam said.

"It could help avoid unnecessary referrals to public hospitals, potentially reduce waiting periods for patients and enable ophthalmologists to focus on patients needing treatment and surgery.

"It could also help reduce the financial impact of diabetes on the Australian economy, which is estimated to cost up to $14 billion a year."

"There are so many areas in healthcare that benefit from AI. The opportunities are endless. AI will redefine healthcare in every single area to improve lives. I am sure we will soon get rid of cancer and give sight to blind people," says Luke, ending the discussion.

As AI advances, how will regulators ensure patient safety while encouraging innovation? This is a dilemma policymakers across the world juggle today.

For decades, the healthcare industry has used AI to increase efficiency and enhance patient care. The technology sector is also partnering with healthcare providers to develop algorithms to improve the current diagnostic practices, from reactive to proactive diagnosis. Machine learning, Neural Networks, Natural Language Processing and Deep Learning are branches of AI commonly used in healthcare applications.

With the rise of telehealth driven by the COVID-19 pandemic, health professionals predict that AI technology will continue to drive innovation to prevent and treat illnesses. Whilst it can improve the quality of care for patients, it raises other questions like data security and privacy. Although the healthcare sector is highly regulated, no regulations target the use of AI in the industry yet. Several countries have proposed regulations addressing the use of AI, but no regulation has been adopted.

The proposed regulation of the use of AI by the EU is the most advanced of all. It includes a procedure for new AI products entering the market and imposes heightened standards for applications of AI that are considered "high-risk". The EU framework provides examples of "high-risk" applications that are related to healthcare such as the use of AI to triage patients for emergency care. The proposed EU regulation strikes a balance between patient safety and innovation in the industry.

In the US, the Food and Drug Administration (FDA) and the US Department of Health and Human Services (HHS) have begun to develop guidelines on the use of AI in the health industry. In 2019, the FDA published a discussion paper covering a proposed regulatory framework for AI-based software as a medical device. FDA assures its commitment to safe and secure AI-based software as medical devices.

The US and EU aren't the only entities that attempt to monitor and govern the use of AI in healthcare. Countries such as China, Japan and South Korea have also released guidelines and proposals to ensure patient safety. In June 2021, the World Health Organization (WHO) issued a report on the use of AI in healthcare and offered six guiding principles for AI regulation: protecting autonomy, promoting safety, ensuring transparency, fostering responsibility, ensuring equity and promoting sustainable AI.

In Australia, the Therapeutic Goods Act (TGA) regulates software-based medical devices, including software that functions as a medical device in its own right and software that controls or interacts with a medical device. In February 2021, the TGA implemented reforms to the regulation of software-based medical devices, including new classification rules for software-based medical devices according to their potential to cause harm through the provision of incorrect information.

One of the main regulatory hurdles with the registration of AI is that it is fluid whereas the TGA review of medical devices is currently based upon a pre-market product at a fixed period of time. The traditional framework of medical device regulation is not designed for AI. Whilst the current regulation in Australia may be non-specific, policymakers are working collaboratively to bring the Australian legislation to the global standards.

Whilst the regulatory action is forthcoming, governments across the world need to revisit the current healthcare model that is fundamentally skewed to treatment care rather than preventative care. Even though citizens are more focused on wellbeing, health systems are geared towards treating illnesses in a way not suited for AI-driven preventative healthcare. This is a challenge that remains to be addressed globally to reap the benefits of technology.

Lesson 5:

Who owns my face?

Dr Joseph J. Atick, the father of commercial facial recognition, explained the discovery to the *MIT Tech Review.*[3]

Dr Joseph J. Atick is considered the "founding father of facial recognition and the biometric industry". After finishing his PhD in Mathematics, he and his collaborators continued working on mathematical algorithms for how the human brain recognises faces.

Eventually, they were able to produce an algorithm that was able to identify faces. In one instance, Dr Atick stepped outside the lab to go to the washroom while the code was compiling. When re-entering the room, the algorithm had finished compiling and was already running. It identified him: "I see Joseph." As others entered the lab, it successfully identified them too. So that was the beginnings of commercial facial recognition in 1994.

3 - J. Strong, A. Green & E. Cillekens, 13 October 2021, In Machines We Trust: "I Was There When: Facial Recognition was Commercialized", podcast, *MIT Technology Review*, viewed May 2022, https://podcasts.apple.com/au/podcast/i-was-there-when-facial-recognition-was-commercialized/id1523584878?i=1000538416500

Subsequently, facial recognition and the use of such technology have become commonplace. It is on most of the devices we use, the phone, the computer, most apps, and some public places like airports, roads, and stations. There are billions of photos captured and stored by the users of technology.

"One wonders who owns the face now," says Indie.

Facial recognition is one of the frontrunner applications of AI. It is one of the advanced forms of biometric authentication capable of identifying and verifying a person using facial features in an image or video from a database.

Facial recognition uses AI algorithms to detect human faces from the background. The algorithm typically starts by searching for human eyes, followed by eyebrows, nose, mouth, nostrils, and iris.

Facial recognition software reads the geometry of your face. Key factors include the distance between the eyes and the distance from forehead to chin. The software identifies facial landmarks that are key to distinguishing your face. These features cannot be altered with surgery. Once all the facial features are captured, additional validations using large datasets containing both positive and negative images confirm that it is a human face.

Some of the standard techniques used for facial recognition are feature-based, appearance-based, knowledge-based and template matching. Feature-based methods rely on features such as eyes or nose to detect a face. The outcomes of this method could vary based on noise and light.

Appearance-based methods use statistical analysis and machine learning to match the characteristics of face images.

In a knowledge-based approach, a face is recognised based on predefined rules. It could be challenging considering the efforts needed to make such definitions.

In contrast, template-matching methods compare images with previously-stored face patterns or features and correlate the results to detect a face. However, this method fails to address variations in scale, pose and shape.

Indie reads from her notes prepared during the week based on her extensive research.

"One of the critical uses of facial recognition technology is face ID on our phones. As you know, most of our digital lives are built around phones and other devices. It means we have to protect our digital data and access to devices. Facial recognition is deemed one of the most secure methods for identifying people.

"As one of the device manufacturers has outlined, the technology-enabled face ID is one of the most advanced hardware and software combinations. The advanced camera captures accurate face data by projecting and analysing thousands of invisible dots to create a deep map of the face. The photo gets converted to a mathematical representation. This is stored in the device itself. Each time the face is captured, it is compared against the stored mathematical data.

"Face ID is smart. It can recognise the changes in your appearance, such as wearing makeup or growing facial hair."

"Mum, is your face recognition unique to you?" asks Luke.

"There is a one in a million chance of a similar face ID, which is a minuscule probability," answers Indie. "So, it is not unique but much safer than passwords or codes."

"Well, can someone download your face ID from the phone?" asks Luke.

"The face ID is not downloadable. It is safely stored on your phone. The phone manufacturer doesn't even save it elsewhere, as they claim. So, it is protected inside the phone."

"This is such clever use of technology to protect ourselves. We should enable face ID and biometrics on all the important apps we have. Do you remember we covered that before? We covered this in the discussion on how to protect our wealth. We also highlighted that if our service provider doesn't use biometrics or Multi-Factor Authentication, it is time to find a more secure service provider."

"Thanks so much for the insightful discussion, Mum," says Lizzy. "But we would like to move on to a more sophisticated area. Why don't we focus on facial recognition in maintaining law and order?"

"That sounds great, Lizzy. Why don't you take charge of the conversation next week?" Everyone encourages Lizzy to lead the conversation next week.

•

It is a rainy day in Melbourne. It has been a wet week for Melburnians. People want to get home on such days and be indoors. The WILL family members are tired after school, sports and work but are very excited about the Wednesday evening.

The two children are making dinner. Lizzy is in charge of making the table arrangements, and Luke is preparing the food. He has become a good baker, and tonight, it is a Luke special. William and Indie are already at the table having a glass of wine.

Lizzy is reading something on the computer. It is her time to lead the discussion on facial recognition, particularly in law enforcement. She is set to lead the conversation. "Let me light the candles," says Lizzy, kickstarting another AI night.

It is raining outside. But the dining area looks beautiful with the elegantly arranged table and the candles that provide much-needed light.

"I researched the companies that specialise in facial recognition. There are so many that focus on this area. I found some of them to be not so useful. But, I found a few organisations that seem to be doing fabulous, so I focused on these companies for my research."

Lizzy mentions that many organisations use facial recognition technologies to support law enforcement in various parts of the world. The primary purpose of these organisations is to identify individuals based on advanced facial recognition technologies to support the investigations of local and federal agencies. With the rise of digital crime, this is an excellent use of technology. We have to make sure that these technologies don't fall into the hands of criminals and the governments known for using power against their people disproportionately. We have had enough examples of this in the past. It is a powerful weapon against innocent citizens.

One such company that specialises in facial recognition technologies outlines on its website that they are dedicated to innovating and providing the most cutting-edge technology to law enforcement to investigate crimes, enhance public safety and provide justice to victims. It sounds very promising and interesting despite the opportunities to misuse it against people.

In addition, they claim that the platform, powered by facial recognition technology, includes the largest known database of over 10 billion facial images sourced from public-only web sources, including news media, mugshot websites, public social media, and other open sources. That's a massive collection of faces. They use social media platforms to identify faces of interest to provide identification information to law enforcement agencies.

"It sounds promising. However, it begs one question," says Lizzy. "Who owns your face? Have we provided approval to use our faces in this manner? I have a lot of social media accounts, and none of them specifically informed me that my photos are used this way. I am sure they have taken my consent using fine prints, the so-called terms and conditions. There are so many terms and conditions present, and most of us click accept without even reading any of the information.

"Anyway, let's look at AI and its use, for good and evil. The evil side perhaps causes unintended consequences of the use of technology. The slogan of companies that provide such technologies highlights civil liberty and justice repeatedly. It sounds as if they provide these services to protect us."

Lizzy relates the following story. One day, law enforcement stopped a person crossing the US–Mexico border. The person possessed drugs, so she was taken into custody immediately by the border security. She was a frequent traveller who had crossed the border multiple times over the last few months without any interception. It was somewhat surprising and embarrassing to the agencies that she wasn't caught earlier. So they decided to have a proper investigation.

Upon enquiry, she showed a social media page that recruited her. The page had a picture of the person who recruited her. It provided much-needed information for the investigating officers. They used facial recognition technology to investigate the masterminds behind the cartel. The officials ran a scan of the photo through this particular system which has over 10 billion photos, and it identified the actual name with all the details of the masterminds. This gave the officials the first level of information to complement other evidence.

Once they confirmed beyond doubt that they had identified the suspect accurately, they arrested the person in a country club a few miles away from the border. The agencies know that masterminds hide behind fake names and use vulnerable individuals in society to do the risky transportation. They usually end up in jails while the masterminds escape corrective action when caught. This system eliminated countless hours of investigative work and the potential to miss the suspect's capture over time.

"Law and order seem like a great example of such technology to prevent harm to innocent people," says Lizzy. "Suppose we use intelligent systems like these for societal challenges like drug trafficking, theft, murder and other criminal activities. In these cases, we should be able to find criminals faster and make society a much better place. The community needs such technologies to work as the first line of defence – we are exposed to immeasurable pain today."

"What if they get the identification wrong? Have you found any examples of where algorithms make mistakes?" asks Luke who wants to know if Lizzy had done her research correctly.

"I have two stories on that, Luke," answers Lizzy. "Let me begin with one such example where a person was named incorrectly as a suspect in a case."

"On 21st April 2019, Easter Sunday, three churches and three commercial hotels in Sri Lanka were targeted in a series of suicide attacks by suspected Islam militants. It was a huge surprise to many who watched in horror what was unfolding in Sri Lanka on international media – breaking news at the time. It unfolded similarly to 9/11. The chaos was rolling on everywhere, particularly in Sri Lanka's capital Colombo and other suburban areas in the South Asian nation. The country was under siege for a day or two, and people didn't know what would happen next.

"It caused massive loss of lives and injuries. A total of approximately 300 people were killed, including 45 foreign nationals. A massive number of innocent people were also injured. Those innocent people died in vain. Some were killed at churches, and others met their fate in popular hotels while having breakfast.

"This had far-reaching consequences. The tragedy devastated the Sri Lankan economy, which relied heavily on tourism. The tourists stopped coming to Sri Lanka in fear of more attacks by terrorists. Together with COVID, the Easter Bombing and a corrupt government, the Sri Lankan economy plunged into the worst downturn in the history of Sri Lanka a couple of years later.

"The Sri Lankan law enforcement agencies were quick to identify the mastermind behind the attacks. On the following Thursday, police issued a notice with the names and photos of six people. That included three men and three women who were wanted for questioning in connection with the Easter attacks that killed about 300 people. One of the women listed on the notice was identified as Abdul Cader Fathima Khadhiya, accompanied by a photo of a woman in a headscarf. But the image, in fact, showed someone else. It was Amara Majeed, an American Muslim with Sri Lankan heritage, the local and international media reported subsequently.

"Unfortunately, the agencies had identified her incorrectly. The photos went viral. This person was living in the US, and she woke up to a barrage of hate messages on her social media. The ordeal must have been a horrifying experience for the person concerned.

"How do you feel waking up to having been identified as a suspect in an international bombing? It can't be explainable. It is a nightmare for anyone.

"The agencies quickly identified the error and posted an apology. Even though not officially confirmed, it was suspected that an algorithm had incorrectly identified the wrong photo for publication. It was great that the agencies corrected the mistake quickly. But, in this day and age where news travels around the world so fast, it must have been a harrowing ordeal for the innocent victim."

"That's a very relevant story, Lizzy," says Indie. "Well done for your research on this topic and identifying real stories you can share with us. You have done a great job." Indie applauds her daughter for a well-balanced and thorough presentation on the topic. The others acknowledge the value of real-life stories.

"Well, facial recognition technology has been put into great use in airports," Lizzy continues. "You may remember our last trip overseas. We simply scanned our passports through the immigration gate that opened the gate with a green tick on the screen. That was pretty awesome."

"If I recall correctly, we were to line up for passport verification by an immigration officer before these technologies were installed at major airports globally," says Indie. "I remember very long lines at immigration before the automation. We, as travellers, have benefitted immensely from the technology. I think the number of travellers has increased over the years due to the adoption of cutting-edge technologies. Technology has made travel so easy and enjoyable. The days of travel without border protection agents looking over your shoulders are soon a reality. It is already in place in some of the major airports globally. Seamless travel will be the norm across other parts of the world where technology penetration is slower as well.

"I remember being questioned at an airport for an extended time when we travelled a few years ago," says Indie, pointing out an annoying experience on a previous trip. "It felt as if we were treated like smugglers. They could have done that seamlessly."

William adds his perspective on the unpleasant experience. "I am delighted that most of these unnecessary things will be automated."

"Don't call them unnecessary, Dad," says Lizzy. "They are necessary things done by humans. However, AI and new technologies have made it a seamless experience for travellers."

"How does facial recognition technology work in airports?" questions Luke.

"First, the individual places their passport on the scanner," says Lizzy. "The system instructs travellers to look at a camera next. The camera takes a live photo of the person holding the passport. Biometric scans attempt to match the stored passport image with the live photo captured by onsite cameras. The algorithm compares your live photo with a stored mathematical representation generated from images on your passport. If the two photos match, then the system recognises it as a valid entry. The gate opens once the photo is matched properly. If there is any doubt, the system flags a manual check that an immigration officer will conduct."

"Has this amounted to a lot of job losses?" Indie asks William.

"Probably," William answers. "But most of the people have been trained in other jobs and redeployed. There was some fear at the dawn of the century that computers and robots would take over our jobs soon. But it has been proved that computers and robots will take over jobs with repetition and leave most of the exciting jobs to humans."

"The system in airports works very similarly to the design on our phones," says Lizzy. "There are other places where this could be easily used, such as hospitals, supermarkets, and train stations. They could be installed in public areas for the safety of people.

"There was a trial in the city of New York recently which attracted intense public debates. Human rights organisations were concerned that the New York Police Department could track people in central areas of New York, speculating that it mainly targets areas of people of colour in the city. Volunteers worldwide participated in a tagging project and identified more than 15,000 cameras in these areas.

"You are never anonymous. Whether you are attending a protest, walking down streets, or going to a nearby store, your face can be tracked by thousands of cameras using facial recognition technologies across the city. The NYPD has claimed to have tracked more than 22,000 persons since 2017, the vast majority of them in 2019. The international human rights organisations and media accuse the NYPD of using technology to track citizens.

"Furthermore, these technologies were used in 2020 to identify and track participants of Black Lives Matter protests held in the city.

"Human rights activists suggest that facial recognition can and is used by states to intentionally target certain individuals or groups of people based on ethnicity, race and gender, without individualised reasonable suspicion of criminal wrongdoing.

"New York is not the only city being accused of tracking its citizens. A search on Google highlights that almost all the countries in the world use facial recognition technologies for the safety of their citizens. There are similar services used in our country as well," says Lizzy with surprise.

"I am not surprised," says Luke. "Let me highlight an Australian example."

Luke mentions that in an effort to limit the spread of COVID-19 in Australia, several services powered by a private firm have been introduced by the Western Australia Police Force.

First, QR codes to manage lockdown restrictions, then travel permits during the pandemic. Some of the early starters in what became a digital technology rush to manage the healthcare challenge have led to many new innovating solutions. A service provider that provided QR code-based travel permit system to manage travel restrictions in Western Australia has evolved into a comprehensive platform that supported the elimination strategy of the state.

It includes a pass, a border management solution used to manage all travel into W.A., and the app that has helped more than 95,000 people quarantine at home in W.A. since September 2020. The app uses facial recognition to confirm the identity of people in quarantine and location services to verify their location. A full-scale quarantine management solution saves significant police resources and is an Australian first. The company claims all this proudly. The media also mentioned that there was a similar trial in Victoria and New South Wales. Political leaders have publicly dismissed these.

Most governments are flocking to these technologies to track COVID-infected individuals. The debate is gathering momentum that governments under the guise of COVID management are rapidly deploying technologies to track the movement of citizens. These technologies are widely used across the world. In some cases, they have been used to protect society. But, there are instances of mistakes and misapplications.

It will be a phenomenal area to watch in the coming years as technological solutions become commonplace in our workplaces, public areas and even on our phones.

Indie asks, "How about in schools? Are there cameras in schools?"

"It is a place where the technology could be used effectively. Even though these technologies are not commonly found in schools and universities, cameras are already in place in most education institutes," answers William. "It is a matter of using facial recognition software if the cameras exist. So, there is a possibility that facial recognition technologies may penetrate schools, libraries and universities.

"Institutions can easily use these technologies to record attendance in schools and universities. It may require approval from teachers and parents. Many schools don't use any of these technologies today. But, it is a potential future application. We will be able to reduce students' and teachers' poor attendance through real-time reporting and monitoring. Upon detecting an absence, the system will immediately notify the parent."

William continues to elaborate on the use of such technologies in schools. Privacy issues must be evaluated before implementing advanced technologies in educational institutions such as schools, libraries and universities. Privacy and data protection are two critical things that should be debated before implementing a wide-reaching policy. Luke and Lizzy agree with a thumbs up. There was recently a trial project in some of the schools in Victoria, Australia. It received widespread feedback from various groups. Most schools decided not to participate in the trial due to privacy concerns.

Lizzy poses a hypothetical situation. "Let's imagine getting the learning and development technology previously discussed coupled with the facial recognition technology. It would be a fantastic place to walk into. As soon as you go through the door, the system will identify you and keep the computer on with the right learning program for that particular day based on your mood."

"I want to suggest something for the coming weekend. We are going to the F1 Grand Prix. Dad has bought tickets for us. Shall we take a count of public cameras in places on your journey to Albert Park?" suggests Luke. They all agree to use the racing weekend to further educate themselves on the subject of facial recognition technologies in their hometown.

That is the end of another fabulous AI night at WILL family home, thanks to Lizzy for the amazing research.

As you have read a few chapters of the book, have you decided to orchestrate a similar concept to WILL's AI night at your place? It is time to give it a chance to provide a foundation for ongoing learning. You may start small and test out how it could work for you. Find the right recipe for success. Don't fall behind. It is not too late to board the AI train as it is just gaining momentum.

•

It is the Sunday morning of the Grand Prix. There is a huge buzz for the race. Luke is a huge fan of car racing.

"Let the journey begins. As agreed on Wednesday night, we have a challenge today. We need to identify cameras in use along our trip to the park," Luke reminds everyone of the agreement.

"Don't worry; we haven't forgotten," says Lizzy. "Let's start from the house. We have installed security cameras in our house for our own safety and security. That means these cameras record our movement. One advantage is they are not in public space. Most of the houses along the way have cameras. It is excellent for the safety of the neighbourhood."

They set off the journey in the family car. The first stop is the train station. They park the car at the station and take a train to the city since it is a hassle to drive to the city area when these events are on.

"One, two, three, four ... There are six cameras, mostly at intersections," Lizzy laments.

"We have been captured in six different places already," adds William.

After parking the car at the station, they walked to the train platform.

There are cameras in train stations for the safety of passengers. Stations, including car parks, have surveillance cameras. They have been in place for years, probably more than ten years old. Consequently, passengers are under surveillance throughout the city as train carriages also have cameras in them for the safety of patrons.

"Once we get off the train, we need to take a tram to the area," says William.

"Trams also have surveillance in place. From the train station to the tram stop near the park, we are under surveillance. That's probably more than 90% of our journey today. How often do we think about these technologies every day?" says Indie.

"Not very often, Mum," answers Luke. "Everyone is captured on those safety cameras when we are inside the park. In a way, most of our activities outside the house are recorded somewhere on a camera. Surveillance is fantastic for the safety of people in cities like ours. It assumes citizen safety but that still leaves the question of privacy."

The challenge is a practical way to understand the impact of technology on everyday life that shows how prevalent the technologies are in our society.

•

In what could be the first recorded case of wrongful arrest due to a faulty facial recognition match, a man was arrested for a crime he didn't commit. As many media outlets reported, including *The New York Times*, the arrest happened in Michigan, US.

A man named Robert Julian-Borchak Williams was in his office when he received a phone call from the police department asking him to surrender himself. He first thought that it was a prank call. That afternoon, when he drove into his home after work, a police car pulled up and blocked him from behind. The police officers got off and handcuffed him on his front lawn in front of his wife and two daughters. It was a devastating ordeal for the family of four. The police didn't say why he was arrested. They just simply showed a warrant with a photo of him on it. Police whisked him away without telling his wife where he was taken to.

The police drove him to a detention centre and took a mugshot, fingerprints, and DNA as part of the process. A day later, two detectives took him out and started interrogating him. They started questioning about his visits to an upmarket boutique shop. "When did you last go to the shop?" asked the detectives. He responded by saying that he and his wife went to see it when it first opened up years ago.

The detectives showed him the first evidence on a piece of paper images from a surveillance camera – a man dressed in black wearing a red cap in front of a watch display. An expensive watch had been shoplifted. "Is this you?" asked the detectives.

The second piece of information was a close-up photo of the man. Robert knew it wasn't him. He picked it up, held it closer to his face and said, "Is this me? This is not me." All black men do not look alike. He knew he hadn't committed the crime. However, he didn't know that an algorithm had wrongfully identified him. That Friday, he was about to celebrate his 42nd birthday.

In Robert's recollection, once he showed the detectives that his face didn't match the close-up on the second paper, the detectives leaned back on the chair and said, "I think the computer got it wrong."

They then handed him the third piece of paper, which he showed again that it wasn't him. He asked the detectives if he could go home. "Unfortunately not" was the answer he got.

He was kept until Friday and was released on a $1,000 bond. His family contacted attorneys in the area, and most of them requested exorbitant charges assuming he committed the crimes. An organisation interested in facial recognition technologies took up the case for him. Two weeks after his arrest, he appeared in county court. The prosecutor moved to dismiss when the case was called, but "without prejudice," meaning he could later be charged again.

A spokesperson for the prosecutor mentioned that a second witness was at the shop at the time. If this person identified him later, the prosecutor's office might decide to charge him later.

Later, the prosecutor's office informed him that he could have the case and his fingerprint data expunged. He and his wife have not talked to their neighbours about what happened. They wondered whether they need to put their daughters into therapy. His boss advised him not to tell anyone at work. "My mother doesn't know about it. It's not something I'm proud of," Robert said. "It's humiliating."

It is a case where humans have misused the technology – humans have not used common sense with the technology. The system should be complementary to the investigation rather than making conclusive decisions. Perhaps, the users have entirely misunderstood the technology solution, thus relying solely on judging someone's character.

The research by Pew Research Center which is a nonpartisan fact tank says that 42 federal agencies that employ law enforcement officers have used facial recognition technology in one form or another. Some recent studies by renowned organisations have found that while the technology performs relatively well on some ethnic groups, the results are less accurate for other demographics due to a lack of diversity in the images used to develop the underlying databases.

We have to be conscious of these biases. Algorithms can have an inherent bias, so such technologies should be carefully implemented and used. The Pew Research Center study also highlights that Americans have mixed views on whether the use of facial recognition technology will make policing fairer.[4]

The debate primarily focuses on how technology is being used rather than the technology itself. There are moral questions on this topic. Do governments have the right to record and use citizens' faces? That is the fundamental question. There are two sides to this question.

The school of thought that are in favour of the technology, such as governments and companies that develop these solutions, argue that this protects human liberties more effectively. The areas such as law enforcement, automation of processes, reduction of fraud and theft are examples where most commercialised solutions have been developed over the last 20 years.

On the other side, many cases like wrongful identification lead to more harm for individuals even though the reported incidents are low. The governments and policymakers have a massive role to play in making balanced policy decisions on the proper use of technologies in society. However, policymaking usually lags a few years behind technological advancement, opening the door for manipulation and individual interpretation.

The critically missing part of the debate is the involvement of the public. There is a lack of understanding among the people about technology. So this has not become a national debate. We have to educate more people on AI and technologies. The more educated the public is, the more advanced the national discussion becomes. That is the duty of all of us. We have a moral responsibility to educate as many as possible on the subjects that advance the human rights agenda.

---

4 - L. Raine et al, 17 March 2022, "Public more likely to see facial recognition use by police as good, rather than bad for society", Pew Research Center, viewed May 2022, https://www.pewresearch.org/internet/2022/03/17/public-more-likely-to-see-facial-recognition-use-by-police-as-good-rather-than-bad-for-society/

"Charity begins at home," says William. "That's why we started the AI night at our place."

"Well, is there an answer for who owns the face?" says Luke, reminding everyone what the original question was.

"In my mind, I own my face," answers Indie. "It is part of me as a human being. It is my asset. It unlocks security and safety. Moreover, I am identified by it. I don't think there is any doubt about who owns the face."

The simple answer is that the individual owns it.

The fundamental question is who uses it and the mechanisms to get approval from the owner to use it, whatever the situation may be. We don't understand how and where our faces are used as a society. There is very little understanding of social media users who share the bulk of images used in these technological solutions. We must raise these challenges of the 21st century more often. I don't think the political and societal structures are bringing them up more often.

It is ultimately down to all of us. As members of the broader society, we should first understand them, then make others understand.

In a move to simplify and rationalise the identity services in Australia, the federal government has introduced Australia's National Identity Security Strategy, which provides high-level guiding principles to guide identity security initiatives. Australian Government's ID Match provides comparison services for personal information against government documents such as passports, drivers' licenses and birth certificates. The Document Verification Service (DVS) checks whether the biographic information on your identity document matches the original record.

In addition, the Face Verification Service (FVS) compares your photo against the image used on your identity documents, usually with your consent. Furthermore, the Face Identification Service (FIS) compares a person's photo with other photos held in government records to identify them or detect multiple fake identities. These government services provide a trusted and secure verification mechanism for proper identification and authentication. At the time of writing, both federal and state governments are working closely to create a centralised National Driver License Facial Recognition Solution. The Department of Home Affairs manages this system on behalf of all states and territories.

Similarly, federal and state legislators are working on introducing legislation to effectively manage Facial Recognition Technologies (FRT) in other jurisdictions globally. EU's proposed AI Act also identifies Facial Recognition Technologies as high-risk and unacceptable risk items within the risk-based framework, which will require specific actions by developers of such technologies.

FRT is a complex area that requires a multi-facet approach to regulation. We will see such efforts within the next few years as the technology becomes more mature.

Lesson 6:

Fakes are getting worse

Luke reads the following disturbing story: "My son and a few of his friends decided to go on a holiday. It was his first holiday without the family. He had been dreaming of this for many years, a day he finally becomes an adult and can do many things his way. As parents, we encourage our children to be independent and courageous. He didn't have much opposition from us as parents. However, we only had one condition. Mum wanted him to do it only after his high school exams, and she also emphasised that he shares details and provides an update regularly when he is out with his friends. That was a small compromise for him.

"My son is a very responsible child. That's how we brought him up. He shared much of what he was doing, mainly with my wife. They had a special bond. They were like best friends as she treated him like her younger brother she never had. She lost her younger brother at a young age. I guess deep down she probably had a fear about her son.

"He had a big group of friends who went to the same high school. They were friends from a very young age. They played sports together and spent most of the weekends together. So we knew them very well. They were like an extended family. We were also good friends with their parents as well. We celebrated many things together, birthdays, anniversaries and Christmas. When he got to his senior year in high school, we knew that the group had become responsible young men.

"They had been planning for a trip for months. It was an overseas trip for a couple of weeks. They had booked everything well in advance as a large group of travellers. I helped them with things like finding good hotels and cheap flights. Ultimately, they collectively decided where, when and how they will spend the well-earned holiday.

"Finally, the day arrived. It was probably the most anticipated and exciting trip of their lives. We all went to the airport to drop them off. Some of the grandparents were also present. The kids were super excited. They were hugging each other and showing their excitement in front of everyone. We all knew that they would be very responsible. There wasn't any doubt about it. After taking a few photos and saying bye to us, they went through to the departing lounge. We waited for another at least 30 minutes before heading back home.

"I can't recall talking much inside the car on the way home. My wife was in deep thought, and I concentrated on driving. We felt something was missing from our lives. After all, it was a two-week holiday. I tried to comfort her, but she was unusually quiet for a day or two.

"We received a call from him; he conveyed that they safely landed and arrived at the resort on time. We could hear kids enjoying in the background, screams, the sound of water, music and big laughs. It was time for them to have good fun.

"My wife and I decided to get away for the weekend as the house without our son felt like a ghost town. We both wanted to avoid it as much as we could. He would be back before the next weekend, and we had a busy week ahead with both of us having to meet work deadlines that seem pretty daunting.

"My wife continued to get updates from our son every day, mostly in the evenings, and she was not happy about the once-a-day update. I guessed she expected him to call every two hours. After all, that's how most mothers are. They are always worried about loved ones. I didn't care much about updates as long as I knew he was safe. I knew deep down he would be fine, and his updates to my wife were more important as I didn't want her to worry about him too much.

"We went to the place we booked as planned on Saturday morning. It was a calming and beautiful countryside location with a nice orchard. We both love places like that rather than busy beach hotels. We had heard about this beautiful place but never got the chance to spend time previously.

"We were having lunch when I received a call from an unknown number. I first declined the call. I don't answer calls from unknown numbers as most are scam callers or sales calls. Even though I didn't answer the call, I was hesitant, not knowing what it was. I thought to myself, *If the call comes through again, I will answer next time.* So, I continued with the breakfast. My wife was unaware of the call and kept taking selfies."

"The call came through again. This time I answered. There was a deep voice down the other end. I knew it wasn't someone I had spoken to before. I felt uncomfortable."

"'Hello, we have your son in our custody, he is safe, and we will release him upon receipt of $100,000. You don't have much time. It has to be done in the next hour, and I will get my lawyer to give you a call soon.' He hung up the call soon after. I couldn't even say a word. I was stunned and didn't know what to do. Before I shared what happened with my wife, another call was again from an unknown number.

"'Hello, I am the lawyer, and your son is in safe hands. The only thing is you have to wire $100,000. Let me first share your son's voice, and I will send you a video of him after the call. Don't take this lightly.' He paused and played a recording in the background.

"'Hi Dad, I am okay, and we all are ok. Sorry, we went outside the resort last night. They captured us in a remote area. Please give the amount they want, and we will be free. They have treated us very well so far. I love you both.' That was a recorded voice of my son. That was his voice. Even the little doubt I had before was gone. I knew he was in danger. The lawyer said he would send the video of our son soon so that we know he is safe. He also insisted that we have to pay the amount within the next 30 minutes.

"That was something I never expected. My son is in the hands of a kidnapper. I have only 30 minutes to fulfill their demands. My head started spinning. However, I decided to handle the situation carefully.

"I turned to my wife and shared what I was informed. First, she was in shock. She almost fell over. She started crying loudly. I received a message on my phone. It was a video link. I clicked on it, and a video played. It was the same message but with my son in the video this time. It looked authentic and genuine. I showed the video to my wife as well, and she was also convinced that it was him who is in the video.

"After struggling for a few minutes, my wife seemed strong and logical. She said to me, 'Hang on, this is not possible. He had posted a video on social media less than an hour ago. Let's call him.' I started doubting the callers.

"We called his mobile phone first. I looked at the international time zone on my phone, which showed 2 am where he was. The call went to his voice message. I tried a few more times but got no answer from him. Then we called a few of his friends. None of them answered. Again, some doubt crept into our minds as phone calls went unanswered.

"Meanwhile, we received another call from the lawyer. He insisted the money be paid within the next 20 minutes. At that point, I knew it was a scam. They were trying to get money transferred as quickly as possible. I begged for another 30 minutes and explained that we were collecting the amount from friends. He insisted that we don't share the incident with law enforcement agencies.

"I tried calling some of the parents. At first, I wanted to give them a call and find out if they had experienced the same ordeal. They all sounded normal. So I decided to share the details with them. Meanwhile, my wife called their hotel. The receptionist didn't know much about the group. However, she agreed to send someone to knock on the doors of rooms where the group stayed. She also tried calling the rooms via the intercom system. It took a while before she could send a room boy to a room. My son had received the message, and he called my wife's number while she was on the phone. He had also tried my number, but I was also on the phone with other parents.

"We both received a message from him, 'What's going on? I received a message saying to call your parents for an emergency. You both seemed busy on the phone, so please call when you are free.'

"So we called him, and he answered immediately. They had had a long night and had just gone to bed. They knew nothing about the callers nor voice and video messages. We were so thrilled that they were safe.

"They came back safely from the trip after that week. We still have a copy of the video, and no one can identify that it was a fake video. It was a terrifying moment in our lives. We didn't receive another call from the lawyer or kidnapper. However, we have a lasting memory of it. I am not sure how to describe that experience even today."

Luke finished reading the article. He had found it from a newsfeed as part of his research on deepfake.

"That's a horrifying story and a scary situation to be in for a parent," says Indie.

Deepfake is a topic that will change the way we see things in the future. It is still not commonplace, but it is gaining momentum. So far, we have discussed wealth, health, jobs and facial recognition. These are not new to AI. So the next topic is a relatively new concept that emerged recently.

Deepfakes are synthetic media in which a person in an existing image or video is replaced with someone else. It uses AI to generate visual or audio content with the potential to deceive people.

Deepfake has gained widespread attention due to its misuse in generating deceiving content like the video in the example above, financial fraud and pornography.

Have you seen the video of Barack Obama calling Donald Trump a 'complete dips—'? It went viral on social media a while ago. It has been recorded using deepfake technologies. This is like photoshopping. We had similar challenges with photoshopping. But deepfakes have gone to another level.

Deepfakes have lately been used for the distribution of fake information. Fake news has a significant impact on society, including elections, people's opinions, and knowledge itself. We had a debate from the 2016 US election on how fake news influenced the election's outcome that year. Some of the social media platforms were called into question on their ability to filter and eliminate fake news. A vigorous debate will have to begin again, this time on deepfakes. There are many more advanced technologies creating fake content using deepfake.

Have you seen the Tom Cruise video on TikTok that has gone viral? It is a deepfake video. There are so many examples of deepfake videos on celebrities, national leaders and iconic figures in society. The videos look authentic.

Deepfake technology can create convincing but entirely fictional photos from scratch. A non-existent journalist who had a profile on LinkedIn and Twitter was probably a deepfake. Another fake on LinkedIn claimed to work at the Center for Strategic and International Studies but is thought to be a deepfake created for a foreign spying operation. Audio can be deepfaked too, to create "voice skins" or "voice clones" of public figures.

Recently, the German energy firm's UK subsidiary chief paid nearly £200,000 into a Hungarian bank account after being phoned by a fraudster who mimicked the German CEO's voice. The company's insurers believe the voice was a deepfake.

"How are they made, Luke?" asks Lizzy.

"It takes a few steps to make a face-swap video. First, you run thousands of face shots of the two people through an AI algorithm. It is called an encoder. The encoder finds and learns similarities between the two faces and reduces them to their shared common features, compressing the images in the process. A second AI algorithm called a decoder is then taught to recover the faces from the compressed images. Because the faces are different, you train one decoder to recover the first person's face and another decoder to recover the second person's face. To perform the face swap, you simply feed encoded images into the 'wrong' decoder. For example, a compressed image of person A's face is fed into the decoder trained on person B. The decoder then reconstructs the face of person B with the expressions and orientation of face A. This is a rigorous process to get it right.

"There are advanced AI techniques that could be used too. It's called a generative adversarial network. It is very technical. In simple terms, you pit two AI algorithms against one another. The first algorithm, known as the generator, is fed random noise and turns it into an image. This synthetic image is then added to a stream of real images that are fed into the second algorithm, known as the discriminator. At first, the synthetic images will look nothing like faces. But repeat the process countless times, with feedback on performance, and the discriminator and generator both improve. Given enough cycles and feedback, the generator will start producing utterly realistic faces of completely non-existent celebrities.

"This demonstrates the power of computing. With more powerful computers, humans can achieve more possibilities with faster machines. A creative process is repeated until the algorithm produces a suitable output."

"Can you make a deepfake on a normal desktop or computer?" asks Lizzy.

"It is hard to make a good deepfake on a standard computer. A powerful high-end computer is required to produce a deepfake. As Moore's Law suggests, there will be more powerful computers in the hands of ordinary users. So, one day it will be possible to create advanced videos on home computers. If created properly, it isn't easy to spot a deepfake. That's the scary part of this.

"Governments, social media companies and organisations have to identify and remove fake messages, which is not a choice but an obligation to them. There is a debate gathering momentum, but it has to be a much broader debate. The aforementioned organisations should invest funds in finding solutions to these issues faster. The response is relatively slow. We have seen this lacklustre response on many topics in the recent past. They need to be more responsive to issues threatening the civic fabric and our collective moral beliefs. Some social media platforms have banned deepfake videos as per recent announcements. They are blocking content that distributes misinformation. This is part of deepfake. But it is not all deepfakes. There is more to be done in this regard.

"With passing time, we will learn how to enforce large organisations take actions swiftly to eradicate this malice. Society anticipates that bad actors and actions are eliminated as soon as possible. It can create a culture where we may have zero trust in information. Compromise of free speech and the right to have information invariably impact democracy. Free speech is one of the fundamental rights, and accurate information sharing is a responsibility, including media organisations.

"Research done by a group of scientists has found that deepfake technology can trick biometric checks. These checks are used in authenticating people using technologies such as facial recognition. Sensity, a security company specialising on deepfake detection, detailed how it managed to circumvent an automated 'liveness test' by employing AI-generated faces. Deepfake has the potential to destroy trusted technologies that are critical for the functioning of the digital ecosystem. This begs swift actions from Big Tech - companies that lead the tech revolution - to collectively squash damaging activities from the digital ecosystem.

"There are three things we have to do about this. First and foremost: education. We have to educate society at large.

There is very minimal understanding of new technologies, particularly emerging technologies. This responsibility sits with all of us.

"Secondly, technological challenges have to be dealt with using technology. While one technology is being created, there should be counter-technology to manage its negative impact. That is primarily the responsibility of the technology sector. There are large and resourceful organisations that should invest in such technologies. The not-for-profit sector and philanthropist organisations also must fund such endeavours as part of the agenda.

"Last but not least, policymakers should act much faster. It takes years for policymakers to implement relevant policies. Policy lag is due primarily to the bureaucracy. The policymaking process should be upgraded to meet 21st-century challenges. Perhaps by leveraging technology is one way of overcoming poor responsiveness. We must see solid progress in the new decade to make society just."

"Are there any areas where we can use deepfake technologies?" asks Indie.

"Greater entertainment, particularly cinema, is one area this could bring excitement. Hollywood has used such techniques even in creating some blockbuster movies. The influence of technology was critical for some great films in the past, such as *Jurassic Park*.

"Deepfake technology facilitates numerous possibilities in the education sector as well. Schools and teachers have been using media, audio, and video in the classroom for quite some time. Deepfakes can help an educator deliver innovative lessons that are far more engaging than traditional visual and media formats. It is yet to be developed as technology is just beginning to emerge."

Luke continues the discussion with a radically different point of view on how it could be leveraged in human rights. "Protecting human rights is another area where deepfake might be useful in the future. Some of the oppressive regimes in the world suppress the free speech of citizens. It leads to many more innocent people becoming victims and the activists who oppose these regimes getting punished. If the activists responsibly use deepfake technologies without abusing them, then it could be a good use of this technology to empower free speech and freedom of man.

"We can use deepfake technology in another area, using icons to spread messages in native languages. Very recently, David Beckham partnered with a health charity to produce a campaign to end malaria. The 'Malaria No More' campaign was broadcasted in nine languages. The campaign was a great example of leveraging technology to advance some of the social battles. In the video, he delivers a powerful message to people worldwide to eradicate malaria. The video looks as if he is speaking in nine different languages. He has a social media following of more than 100 million and has a deep connection in many parts of the world due to his on-field popularity in soccer. Someone like that can move a nation. If the message is in a native language, it is even more potent for his followers."

"Wow, that sounds remarkable. What a great way to use technology," says Indie, impressed by the use of deepfake in one of the most challenging human missions of our time. "We should continue to see the trend as it has made a significant impact worldwide."

"There are other areas we can safely use the technology," says Luke. "We have been using digital voice assistance for some time. Siri, Alexa and Cortana are voice assistants in phones, computers and other gadgets. These assistants are faceless. However, they still feel more like a new invisible robot interface that we have to shout at to promptly get the information we need. It doesn't feel natural. Most humans prefer to speak to a fellow human than a robot. The ability to mimic faces, expressions, and voices is critical in the next generation of virtual human assistants. These so-called virtual humans are slowly but surely entering the mainstream through digital influencers that people interact with, similar to human interactions. And while digital influencers don't respond to you in their own words, they are heralding a future of 'natural' interaction with actual virtual beings.

"So, deepfake tech that is trained with tons of examples of human behaviour could give intelligent assistants the capacity to understand and produce conversation with lots of sophistication. It is an area where we could use the new technology in a meaningful way to improve customer interaction and experience.

"It would be good to have a human face for your voice assistant, Siri or Alexa. We will probably use the assistance more often."

"How about customer experience? We have calls, chats and messages with most of the organisation. It is an invisible conversation," asks Indie. Everyone agrees with the question.

Luke responds, "There are areas where such technologies can be so beneficial for everyday life. It will not be harmful if there is an algorithm with a face that assists us in billing inquiries or complaints. All that we need is a seamless experience. Today, we struggle to connect with humans most of the time due to the long wait time. More often than not, we leave unsatisfied with the outcome of customer service partly due to a lack of human interactions. We are yet to see meaningful transformation in customer experience. There are two frontiers of organisations: those born in the digital age and others that started before that. We mainly struggle with legacy organisations that struggle to modernise their customer interaction tools rapidly.

"A company called Synthesia, an AI video generation tool, allows users to create digital versions of themselves that can narrate PowerPoint presentations, translate speech into multiple languages, and serve as AI avatars. The possibilities are endless. We have to use powerful technologies to improve our lives and not harm humans.

"The best way to fight the harmful use of technology is three-fold, as outlined previously: education, technology and policy. I like to coin this ETP approach.

"We have to educate all humans about these technologies. We have to fight the bad actors with technology, and we need to have proactive policymaking."

eSafety Commissioner, an Australian Government initiative to keep their citizens safer online, recently issued a position statement on deepfakes. While outlining the recent developments and challenges, it outlined a holistic approach to counter the negative impacts of deepfake. Raising awareness, a complaint reporting system, developing educational material and supporting the industry through safety initiatives are policy actions the organisation is implementing in Australia. Similar legislative efforts across most OECD economies will strengthen the existing legislative framework. However, more swift and decisive actions are needed to eliminate the impact, particularly on women, due to revenge porn.

Another AI night at the WILL family home ended on a high note.

The WILL family is embracing the technology. We should step up the momentum of learning AI. No one should be left behind. We can start the process as soon as possible. If the lack of time is the key reason, then you may start small; probably 15-30 minutes per day is more than enough. If you have no family members, then invite a few friends to your place. They might not embrace AI as a subject but you can start with the topic that works. The choice is yours.

Lesson 7:

In the eye of the beholder

There is a sound of a doorbell. Someone unlocks the door. The door opens, and three people enter. "There was so much traffic today, and I had to go to three different places for pickups today," someone says loudly like a complaint. They all sit down on the couch. They seem tired and irritated. There is evening news on the TV. The focus of the news is on the Russian invasion of Ukraine. It's a topic for deep discussion with the world largely sympathising with the predicament of the Ukrainian people.

There are two visitors at WILL's house. They are not total strangers.

Charlotte is one of the visitors. She is 91 years old. She has lived in a retirement village for the last 15 years. She is strong and healthy for her age. She spent her early years in a Melbourne suburb close to the industrial areas. She grew up during the Second World War, and most of the men in her family were war veterans. She married young and had four children. Most women of her era joined the labour force working on farms, in factories and in shipyards as men took part in the War.

Thousands of women of her age joined the workforce in many skilled and unskilled jobs transforming society forever. She loves talking about the War and life afterward.

The second visitor is Elizabeth. She is 70 years old. She grew up in the 1960s, an era she still calls the "best time of her life". The 1960s was one of the most tumultuous and divisive decades in world history. It saw the birth of the civil rights movement, greater moves towards equality for women in the workplace and the beginning of technological advancements – the landing on the moon for the first time, Beatlemania sweeping around the world, and mass ownership of automobiles. Elizabeth has two children, and both live in Australia. Elizabeth lives in a Melbourne suburb with her husband of more than 40 years.

Charlotte and Elizabeth are mother and daughter. Indie is Elizabeth's older child. Together with Lizzy, the four women in the room cover 90 years of history. They are a closely-knit family unit. They represent generations of progress, advancements and change. They are at the WILL's family house for a reason.

William and Luke are surprised by the two visitors. They usually only visit on special occasions such as family get-togethers, birthdays and anniversaries.

Luke whispers to her mother, "What is happening today?"

Indie ignores the question and continues talking with her mother and granny.

"William, let's order food tonight," Indie suggests, pointing to the phone. William starts taking the order from everyone before placing them using a popular delivery app on his phone. "It will take about 45 minutes for the delivery," he informs the visitors.

Lizzy is busy arranging the dinner table. She first assembles a bunch of books on one side of the table. It appears to be in chronological order, older books first. In front of the books are small items; bottles, capsules, and sprays are mainly stacked in one corner. No one questions what is happening, aware that it would be a surprise family activity.

Time is fast approaching 7 pm. "Please listen to me," says Indie. "We have four generations of femininity in the room. I invited mum and nanna to the discussion today. Lizzy and I decided to focus today's conversation on beauty. How is AI advancing and impacting beauty? Before we get to that, I would like to go through the evolution of beauty over four generations through the experiences of all the women in the room. We have gathered all the important materials: photos, albums, and devices on one side and, on the other side, cosmetics and beauty materials from each one of our generations. I am sure you both can contribute to the discussion as well."

"Beauty during the Second World War was non-existent by today's standards," says Charlotte. "Women's role in society was mainly to take care of household activities, and men took most of the work in the workforce. The women's role in the family was to look after the children. Things started to change as the war raged on and men participated in combat. Women stepped up to take over jobs seemingly impossible for women due to a vacuum of manpower. Wartime restrictions impacted everyday life, and cosmetics were the first to go. Most items were rationed. It was after the war that things started to settle down. It was about ten years of a harsh life. The mass media advertised primarily through radios and newspapers. The newspaper advertisements had full-page illustrations of cosmetics. Women wore makeup mostly at special events such as Sunday church, weddings and other parties, which were few and far between. We barely had anything. It was like a luxury."

"Well, during my time, it was very different. The theme was young and beautiful," says Elizabeth. "The Kennedys had such an impact globally. They were the epitome of the family in the 1960s. Television was at the centre of the family room. We learnt most things through advertisements on television. It was like a new invention. Cosmetics became an everyday thing. More women started working, and family income flourished during that time in the western world. We wore cosmetics every day, even during weekends. It was a must-have wearable. We had to look beautiful, which was the message. I had access to more things than Mum possessed in her whole life. It was due to changes in time and fortunes. Technology played a huge part in making people understand the progress."

"I grew up in the 1990s, which was one of the most peaceful times of human history," says Indie. "This era was the golden age of the industrial revolution. As a girl, we started using cosmetics at a very young age, probably influenced by the toys we played with. Television and the internet played a huge part too. We watched TV mostly during the evening. We had the luxury of computers and the internet on our desks. The range increased so much over time – daytime usage to night-time and much more. We wore cosmetics to the school as well. Social media didn't exist at that time, but society was massively embracing beauty. It was like an essential item. I think we were mostly influenced by the role models and toys we played." Indie pauses and looks at Lizzy.

"I don't need to say this, but our generation was mainly influenced by social media. I haven't even seen a radio," says Lizzy. "We are exposed to the world at a click of a button at a young age. There is so much information available. Everyone is very concerned about how you look. There is another aspect to it. You can see how everyone else looks as well. It has a double-whammy impact on us. This stack of bottles with all the vitamins belongs to our generation." She counts ... one, two, three ... "There are more than 25 items in the kitbag of women. Other non-physical things like online filters and apps are common beauty items. So there are more than this in our kitbag today. It is part of you. You carry them everywhere – some inside the bag and others inside devices; that's so much progress since the 1940s."

The photo albums on the table demonstrate the evolution of beauty. On the far right is a phone with social media apps. That's where the world today is. There are many social media apps – some with photos and others with videos. On the far left is a wedding album from the mid-1940s. The images are in black and white. That shows the exposure and extent to which things have changed in the 21st century. Charlotte must have experienced so much during her lifetime. Even though she is not on social media, she has witnessed its extensive use of that through younger generations.

The beauty industry has also evolved so much over the years. It is visible through the role-play of women in the room, each representing a different era in modern history. Nowadays, AI is widely used across different parts of the beauty ecosystem. AI has infiltrated the landscape of beauty similar to most of the other aspects of life.

The topic of beauty has disproportionately impacted women in society more than men. As soon as the word "beauty" is mentioned, we look at everything through a feminine lens. There is no clear definition of beauty. Beauty is clearly in the eye of the beholder. However, there has been so much research on the meaning of beauty, mainly in the recent past. The AI or companies that develop algorithms use some of this scientific work to implement solutions for their consumers. The beauty industry is more focused on the face than other body parts. One wonders why that is the case. While beauty is more centred around the whole body of the human, it is common that facial attractiveness plays a bigger role than other parts of the body.

Evolutionary biologist Gillian Rhodes published a paper in 2006, "The evolutionary psychology of facial beauty", that describes the elements of facial beauty including averageness, symmetry, and sexual dimorphism as good candidates for biologically based standards of beauty.

Averageness is how close is your face to the average morph of all the people in your race or ethnic group. How free are you from basic cosmetic flows? There are proportions. Is your face proportionate? Numbers and mathematical calculations measure this.

Symmetry is a sense of harmonious, beautiful proportion in balance on your face. The paper says symmetric bodies are attractive to many animals, including humans.

Sexual dimorphism is how much you look like a man if you are a man and how much you look like a woman if you are a woman. In other words, how strong your masculine features are for a man and how strong your feminine features are if you are a woman.

Averageness and symmetry are both attractive in male and female faces, with medium to large effect sizes in all cases. Sexual dimorphism is also appealing. Femininity is attractive in female faces and is preferred to averageness. Masculinity is also attractive in male faces, although the effect is smaller than for female faces, and average traits also contribute (independently) to male attractiveness.

She further writes, "We have seen some evidence that attractive traits may signal health, which is an important aspect of mate quality, although the evidence is far from compelling. And we have seen that the way our brains process information also shapes our preferences."

She concludes with a note highlighting areas that need further research and study.

There are many exciting directions for future research. More studies are needed on whether facial attractiveness and its components signal health and other aspects of mate quality. Recently, male facial attractiveness has been linked to genetic heterozygosity at sites involved in immune function. Future studies should determine which components of male attractiveness (masculinity, averageness, symmetry) mediate this link and whether female attractiveness is also linked to heterozygosity at these sites. A more direct test of a link between attractiveness and immunocompetence could also be done by challenging the immune system.

"This is the first time I heard a proper definition of facial beauty. It provides a good foundation for our discussion. I am not sure Mum and Nanna ever knew any of that. For that matter, I hadn't heard a scientific definition of beauty," says William while leaning forward to look at Indie's face with a massive smile.

"This is good grounding for us. We have a scientific understanding of what beauty is. I think it is important to have foundational knowledge on this subject before looking into AI. Beauty is a very delicate subject, and we can easily offend others based on our perception of beauty," says Luke.

"There are three broader topics we would like to cover. We will focus on the current landscape."

» Beauty filter

» Beauty algorithms

» Beauty consulting

"Well, that sounds great," says William. "I would love to share a story about someone who had a significant negative impact on his life due to how he looked. It is a great example of how beauty, particularly the negative side, impacts humans. Let me share that with you all before getting into AI."

William then told the following story.

Norman was a boy who was in my high school. He was from an upper-middle-class family. His father was a lawyer, and his mother was a doctor. He was not particularly talented at anything, in sports or academically. He was an average child. He had two younger sisters. He was a quiet kid in the class, not active and not disruptive. No one had any issues with him. He was very nice and kind to all.

There was one small issue. He had a big nose. It was pretty big, disproportionate to his face. It stood out very clearly, and everyone noticed that.

It was not an issue during primary school for anyone. However, when reaching puberty, it was obvious that his nose was too big for him. He must have felt uncomfortable with that. However, he didn't show any sign of dissatisfaction or concern. Initially, no one cared about it. But at some stage, kids started to find things to annoy other children with. It started mainly with kids who were much more muscular and more athletic, who were dominant athletes in the school. Some of them acted more like big brothers and looked down on other kids.

In Year 8, it was a time when computers and the internet were starting to become common tools at school. There were computers in the lab and students used the IT period to learn computer skills such as programming.

Kids started calling Norman out for having a big nose. His nickname was "Pelican". Initially, it seemed like he didn't care much. Something that started as a whisper became his name after a while. In no time, Norman became Pelican. It was strange for a kid who was very innocent and didn't have much to say about anything in the classroom. Over time, this must have got to him. He started reacting to some of the bullies. He initially asked them to stop calling him names which fell on deaf ears.

He didn't have many friends to defend him either. He was a quiet kid who sat in the middle of the class and barely spoke or said anything in the classroom.

There was no opposition to this behaviour from anyone. School authorities were not aware of what was happening. I don't think Norman dared to complain to anyone. So, the group continued their heckling.

One day, as some of the class was walking up to the IT room, they heard a big laugh. Everyone was curious as to what was happening in the laboratory. As they entered, they were all amused by what had been done. Someone had gotten in early and placed a screensaver on every computer with the world "Pelican", white text on black background bouncing around from left to right and back. No one could miss it. The whole class erupted in laughter.

Norman was so embarrassed but didn't say a word to anyone. He walked with his shoulders down and sat in front of a computer. After that, the IT teacher came to the class, so he didn't notice anything as the computers had been rebooted. It was a prank but very hurtful. I saw something in Norman's eye that day that expressed how he felt about it. A few went to him after the class, but he didn't want to say anything. Norman didn't come to school for a few days after that event.

The following week, he had a band-aid on his nose when he returned to the school. It was usual for boys of that age to have small injuries to their faces due to sports. When asked, he said that he had accidentally cut his nose. The band-aid came off after a few days, and there was a mark on his face. The heckling continued as usual. He was continuously called "Pelican" and there were signs put up in places around the class from time to time.

Kids started noticing unusual behaviour from Norman. Occasionally, he had more cuts to his nose. When asked, his answers alluded to that he was playing sports after school.

One day, a few boys returned to the classroom from the toilet, and they had witnessed something unusual. They had seen Norman do something very irrational. He had jammed his nose to the bathroom door.

When asked about it, he said that he wanted to make the nose smaller. It became a discussion point among the classmates.

A few decided to inform the class teacher. The class teacher summoned Norman to his office.

I have a big nose," he said. "I don't like my nose. Most of the kids call me Pelican. I want to get rid of the extra part that makes me ugly.

The principal summoned all the kids for an assembly. Norman wasn't present. He spoke to the classroom in a very severe way. He didn't punish anyone in front of the class. But he spoke to all the kids individually afterward. Some of them accepted what they did and apologised to the principal. The principal also spoke to Norman's parents and informed them of what had happened.

Norman returned to the class after a few weeks. He seemed normal. However, he went to the toilet during class time which was unusual. One day, a teacher asked a couple of other students to follow him discreetly to the toilet. The two students later returned, surprised. Norman had pulled a small pocket knife and started scraping his nose. He hadn't seen anyone come inside, so he continued. The two students got frightened seeing that so they rushed back to the classroom and informed the teacher. Even though they whispered, the whole classroom heard what they had seen. Norman returned to the class minutes later. The teacher excused the class and returned to the teachers' office. The class was dismissed early.

Later that day, Norman's parents came to pick him up from school. There was a sombre mood in the class. Everyone felt guilty about what transpired. Norman never returned to the school. It was later found out that he had continued to cut his nose until his face was fully damaged.

"When I met Norman years later at a public gathering, I couldn't first recognise him," says William. "He had gone through surgery, probably plastic surgery and looked much different. He came and introduced himself to us and briefly explained what happened after.

We underestimate the impact of beauty on humans. In Norman's case, the lack of beauty caused the ordeal. Beauty is a vital subject that we all should understand well."

People gain confidence through beauty, and some lose confidence due to a lack of beauty. In the 21st century, there is so much help available. AI is one such area we are going to discuss today. But there is more than that. The most important thing is to consult with people you trust. In Norman's case, he didn't discuss his challenges and desires with anyone. Had he done that much earlier, he might not have gone through what we went through.

"That was a great example of the beauty that highlights an unspoken area. Even though it was a sad story, it shares a critical message for all of us," says Indie, applauding William for his insightful story.

Beauty filters

Beauty filter is one area AI has had an impact for some time. It didn't exist on a scale before the adoption of social media. It is a software application that applies an aesthetic effect on photos or videos to enhance the attractiveness of the object.

Some of the features include smoothing skin texture or modifying facial features. In other words, filters touch the three attributes of averageness, symmetry and sexual dimorphism. The filters have been built to adjust the characteristics of a person based on predefined logic.

Most of the filters are inbuilt applications in social media. Other standalone applications provide photo editing functionality. Facetune2 is one such application. As highlighted on their website, it offers various functionalities such as skin toning, eye styling and background lighting, among other things. It claims to have more than 100 million users who use the application for selfies on social media.

There are inbuilt filters in most social media applications. Building filters is a large industry of app developers who make money by offering these capabilities to end-users. These features are commonly found on Instagram, Snapchat, and TikTok. It is a type of augmented reality. Augmented reality, or AR, is widely known as an experience where designers enhance parts of users' physical world with computer-generated input.

Invariably, the end-users of beauty filters are mainly young female users who want to present a better version of themselves on social media.

"I started using filters when I was 12. Everyone in my school uses filters. We were so excited when they were first introduced," says Lizzy. "We had so much fun using various filters such as monkey, pig ears, and other funny filters. But beauty filters are applied very commonly by everyone I know. It has become an essential thing in social media."

As previously mentioned, beauty filters are developed using AI algorithms that are trained by using good and bad photos repeatedly. It enables the algorithm to learn and make decisions.

When a specific beauty filter is selected, the features taught are applied to the image the user has selected. If the photo has a blurred side, then a filter for a smooth complexion adjusts the pixel in the target area to achieve the desired outcome in the image, in this case a smooth effect. A beauty filter must first detect the contours, proportions and individual attributes of the person in the picture before it can superimpose a standardised filter function that fits precisely over the photograph. It is a complex process. Nevertheless, computers are fast enough to do this in a flash.

Beauty filters have expanded beyond social platforms. Recently, Zoom, the meeting and conferencing app, has launched a feature that improves a person's visual appearance. The company describes this as "a softening effect to minimise the visibility of imperfection". It is the start of the introduction to beauty filters for in-office tools.

However, critics argue that these filters have a lasting impact on the end-users, particularly young girls. We desire and look for the perfect appearance. However, that is not possible in the real world. The ideal appearance is achieved in the virtual world through software applications. There are no software applications for real life. When that is not possible in the real world, young users of social filters get frustrated. This will have an impact on self-esteem. This can also lead to rejecting the way we really look, resulting in insecurity and depression. Norman's experience is one such example, even though it was not related to beauty filters. All this may lead to people wanting to do surgeries to look similar to their virtual selves.

There is another aspect of negativity debated globally. Most of the algorithms are built based on caucasian faces. It has a limiting and negative impact on non-caucasian. Beauty can vary based on your race or ethnicity. If algorithms are trained mainly on partial data such as faces of caucasian people, then they are likely to apply the wrong filters for non-caucasian users.

Beauty algorithms

Beauty algorithms are built very similarly to the beauty filters. Algorithm logic is made based on good and bad photos or features. Beauty algorithms are used for personalised product recommendations. These are predominantly cosmetic products. Algorithms recommend try-on products as recommendations, in some cases for free. These products cover skincare, hair products, foundations, lipsticks, hair removals, etc. They operate as platforms with all the recommended brands listed on their sites. It performs the role of a beautician. No human beauty specialists recommend a product such as foundation or lipstick for you – an algorithm takes care of personalisation. The complimentary try-on it offers is no different to what we experience in retail malls. The only difference is AI is doing the job rather than a human in the physical world.

Some of the platforms with AI algorithms have a virtual try-on. The user needs to create an account and upload a photo. The algorithm provides a recommendation for all the different categories. The user can try it on virtually and see how it looks before purchasing any products. AR-powered beauty try-on experiences can help online shoppers discover and experiment with various beauty products and shades online through platforms of their choice.

AI-judged beauty contest is another emerging area for the use of AI. There was an event back in 2016 which was the first International Beauty Contest judged by AI. The organisers claim that more than 5,000 people from 100 countries submitted applications and photos for the contest. Algorithms selected 44 winners. However, there was one glaring mistake as claimed by critics. Out of 44 winners, nearly all were white, a handful was Asian, and only one had dark skin. The entries included people of colour, including large groups from India and Africa.

Whilst there are emerging use cases for AI, the technology developers have a massive role to play – removing bias from their products.

Algorithms are claimed to have racial bias and systematic errors due to the very nature of data used during software development. If the data is only a partial representation, then algorithms will most likely deliver biased outcomes. As outlined in the definition of AI in this book, data is critical for robust AI applications. If AI is based on garbage data, the performance of algorithms will deliver garbage outputs.

It is absolutely critical that developers immediately make appropriate design considerations as we move into a world where algorithms will take over some of the repetitive tasks from humans. Without it, AI will not achieve the potential it is claimed to have.

Beauty consulting

Beauty consulting has been a lucrative industry for many decades. From skincare to cosmetic surgery, they have grown exponentially due to the widespread interest in cosmetics. We are beginning to see the use of technology in this area too. AI has invaded the beauty consulting industry. The day AI algorithm provides the recommendation and carries out the surgery without human intervention is not here yet. However, AI is collaborating with human specialists to offer advice to their clients.

These companies scan human faces and provide a series of recommendations. Some of the suggestions involve cosmetic surgeries. As outlined above, the algorithms are built on scientific principles. The recommendations highlight the variability and solutions thereon to principles of beauty.

William poses a question to Lizzy and Luke. "How popular are they among the younger generation?"

"Well, it is somewhat popular among young women. I think some of my friends were considering a few options. They are teenagers," answers Lizzy.

The surgeries have two facets, essential cosmetic surgeries and optional cosmetic surgeries.

The essential cosmetic surgeries are operations performed to remove a deformity that is either a birth defect or acquired. These operations form part of the crucial healthcare system. The AI technologies will make them more accurate, robust and fast. Consequently, it will improve the productivity of the system.

However, optional surgeries are, as its name suggests, not essential. The individual decides to undertake such procedures to improve their appearance. These raise moral questions. The obvious ethical questions have two opposing views.

One may question the negative impact these surgeries have on the patients themselves. They have to endure the pain and bear the financial cost of such procedures, which are not covered as part of government medicare. These cosmetic surgeries are expensive. Those who are not fortunate enough to afford such treatments will have to bear the pain of social stigma if we continue to push beauty beyond affordability.

However, we have to ensure that bullying cases like Norman's situation don't occur. Those who are in favour of such treatments argue that the improvement in appearance and beauty has a significant impact on one's mental health. As such, these surgeries are essential for a good and healthy life. The proponents argue the case for coverage in such instances under universal healthcare.

Cosmetic surgery is another topic that requires national debate. However, it has to come from responsible organisations that have a stake in women in society. Many interested stakeholders should be involved in making correct policy decisions for the benefit of future generations.

"What a wonderful discussion. Thanks for inviting us to be part of a significant discussion," says William. "We are glad that we could bring back a bit of history to add value to the conversation." Both grandparents raise their glasses to cheer the younger group.

Lesson 8:

Our vehicles

The year was 1939. The war in Europe started that year. The USA didn't get involved until much later. Life was flourishing after the Great Depression in the western world. Jobs were abundant, mainly in factories. The industrial revolution had started. This had put so many automobiles on the road, which hadn't been designed to cope with the unexpected surge of cars on the road. The cars had taken over the carts as a primary mode of transport. Traffic congestion was one of the main issues of governments at the time.

The movies of the 1930s depict the chaos on roads where pedestrians, rickshaws and cars share the same narrow strips in major cities.

There was an exhibition in New York called "Futurama" in 1939. The show promised a future where technology would provide a seamless solution to the traffic problems of the day. An exhibit presented a future model for the world in 20 years. General Motors sponsored this exhibit.

Futurama was a model representing almost every type of terrain in America and depicting how a motorway system could be laid down over the entire country. The layout covered mountains, rivers and lakes, through cities and past towns – never deviating from a direct course and always adhering to the design principles of safety, comfort and speed.

The network of streamlined motorways was designed around a landscape of half a million buildings, a million trees and fifty thousand miniature cars travelling on a fourteen multi-speed lane highway. It depicted a future of self-driving cars where cars would directly communicate with the roads to move safely to the destination. It was magical and might have looked like the perfect solution to the traffic congestion problem at the time.

Nearly 20 years later, in 1958, General Motors produced a working prototype of Futurama. That was probably the birth of autonomous vehicles. Sixty years after the production of the prototype, we are still driving automobiles. There is still a person needed behind the steering wheel.

The automobile industry has become one of the largest in terms of jobs and economic significance. It supports millions of jobs, car drivers, truck drivers, train drivers and all other supporting jobs that maintain people's livelihood. Trucking is one of the leading industries where the truck driver makes a reasonable income to support his or her family.

In the recent past, ride-sharing has penetrated every country and major city, promising to make travel faster and more affordable. The ridesharing industry has transformed how we commute. Some of the brand names in the industry have become words in vernacular English.

How long will it take to achieve the dream of fully autonomous vehicles on roads? It has already been 80 years since the concept in Futurama in 1939. Some expected this to happen by 2020 but we are yet to see fully autonomous vehicles on roads at every doorstep.

AI in vehicles

AI, as previously defined, is a collection of technologies and techniques working together to enable machines to sense, comprehend, learn, and act rationally.

If we apply the definition to the context of automobiles, AI is involved in many parts of a vehicle. It is simply not just one thing.

How do humans use vehicles? Let's understand our automobile journeys. First thing first. We have to know where the destination is. If the destination is a known place or a regular place, we tend to pick up a coffee or breakfast on the way. That decision-making takes place in people's minds. The driver should first conduct a circle check before getting in the car. Once seated, the driver should check visibility, fuel level and other passengers (if there are any) before starting the ignition.

Once the driver starts the car, he or she needs to understand the surroundings. Is it clear to take off? Is the heating or cooling level right? Is the favourite music track playing?

There is so much to decide.

When the vehicle is on the road, there are defined rules and regulations all road users have to adhere to, including traffic lights. There aren't traffic signs on some roads. But we understand the accepted speed limits, such as in built-up areas or on freeways. Some roads and signs may be covered with snow or some unexpected obstructions while driving. Sometimes, the traffic lights are not working due to a fault or an accident.

What about the coffee? We changed decisions halfway through as traffic was heavy and we had to reach the destination quickly. So, we skip the coffee to save time.

Once we safely arrive at the destination, finding a parking space can be a battle. If it is in an office environment, the car park might be full. So, we park near an adjacent area that allows non-resident parking. It is very similar in shopping centres. We need to park closer to the shops, not where parking is available.

So many small things make up a trip. There are so many parts within the vehicle that are used. All of these small parts have to be AI-driven. Once we have all the things working seamlessly, we can have a fully autonomous or driverless car on the roads. That is the ultimate autonomous vehicle.

How safe should autonomous vehicles be?

There are 1.3 million deaths per year due to road accidents. Road accidents are one of the leading killers of people, particularly young people. Everyone is determined to achieve zero accidents on the road. Many ambitious and determined people are working to achieve this goal for humanity. Our collective wisdom will enable us to achieve this dream one day.

William poses a question for the family. "I want to do a small survey among ourselves. From each of you, let's understand if you think autonomous vehicles should be more or less safe than their human-driven counterparts?"

Luke makes an argument. "They should be safer than we have today. That's why we are embracing technology. If they are less safe, then why do we spend billions or even trillions on using them?"

"I completely agree with what Luke said," says Lizzy. "Autonomous vehicles should be much safer. We should aim to achieve zero accidents through innovation. This is our chance to eliminate unnecessary deaths on the road."

"I have experienced road accidents myself. The impact is very severe mentally, even after a small mistake on the road," says Indie. "We humans tend to make mistakes. I want my autonomous car to be fully safe and zero-accident guarantee. Without it, I wouldn't trust the machine to take me around. I want that peace of mind."

"Well, I am like you guys. I want my safety assured. If it is not guaranteed, I might as well drive the manual vehicle," says William. "What's the point of having something that is not trustworthy? I also advocate a zero-accident guarantee."

We expect autonomous vehicles to achieve what we as humans haven't been able to on the road – accident-free driving. They have to be much safer than how we operate them today, which is a gigantic task.

The automakers and innovators are currently working on this challenge. They have to achieve something we have not been able to do. However, it is possible for AI and machines since they are not like humans. They don't fall asleep or get distracted. The key is achieving the optimal level of performance and decision-making at lightning speed.

What is autonomy?

We need to understand the definition of autonomy in this context. There are 5 levels.

Level 1 is having no AI in the car. It is a basic car without AI-driven technologies. The human driver has to do everything in such vehicles. The vast majority of cars fall into this category.

Level 2 is having some level of AI in the car to aid or assist the human driver. It includes but is not limited to features such as speed warnings and emergency braking, among other things. There is still a human driver behind the wheel but there is some assistance for the driver.

Level 3 has adaptive control in the vehicle. For example, when you are on a highway, the car is aware of this and takes over cruise control with your consent. It adjusts the speed based on the traffic and manages the manoeuvring within the same lane. In some situations, the vehicle can take over, but the human driver has to be behind the wheel.

Level 4 is where the vehicle is doing everything automatically. It drives the passengers safely to the destination. There is no human intervention at any stage; however, the human is behind the wheel in case of an unforeseen event when the vehicle requests that a human take over the driving.

Level 5 is where there is no driver at all behind the wheel. There is not even a driver's seat in the vehicles. All the seats are passenger seats. The steering wheel may disappear entirely from the design of the car. The vehicle drives the passengers point-to-point and takes care of parking until they return. The vehicle arrives just in time to pick them up once they return from their engagements. This is probably the ultimate objective of the industry.

The vast majority of vehicles on the road are levels 1 and 2. There is some assistance for the driver, but the human driver takes care of everything. Some vehicle models partially meet Level 3 criteria, such as Tesla cars in commercial production.

Most of the automakers are testing level 4 entry-level cars, and we will soon see some of them out in consumers' hands.

William poses the second question. "Let me ask a clarifying question from all of you. Which level do you want to have or own?"

"Let me be clear, Dad. I am a car enthusiast. I love driving cars. Even though I still can't drive, I want to drive my own car," says Luke. "I may like some of the driver-assisted features in the car. However, I will not be satisfied sitting in a car watching outside through the window. I will be driving my car."

Indie, however, is very enthusiastic about what the future holds in terms of getting an autonomous car for herself. "I am very different to Luke. I love my coffee and reading. I'd take it if I could sit in the car and have 30 minutes for myself. I want a Level-5 car. I can achieve a lot more with 30 minutes for the rest of my life. Bring on, baby! I want it tomorrow and am happy to pay for it. However, I have one condition. It has to be accident-free. The manufacturer should provide a zero-accident guarantee."

Lizzy expresses similar views to her mum. "I am like mum. I want a Level-5 car. I want it to be hassle-free too. I can't be bothered worrying about what's going to happen. I don't want to do anything with driving. Perhaps, that's my time to do studying, chat or write. I can chat with my friends when I am sitting in the car."

"I am either level 3 or 4 like Luke," says William. "I don't hate driving. I am not someone who can do other things inside the car. Driving simulates my brain. I enjoy the challenge every day. It is my morning coffee, in a way. I don't want to lose that either. So I settle down with a human-driven car."

The family is split half-half, with some wanting full automation and others seeing driving as a form of entertainment or mental simulation. This is not a representative sample. However, if we were to learn one thing from the exercise, it would be that we are not going to get rid of human drivers soon on roads.

"What would it take to sit behind and relax in a vehicle? What if the car has a gaming console inside where you can enjoy your favourite game with friends?" Lizzy sarcastically asks her brother.

"That's a great suggestion. If I had something I like inside the vehicle, I'd probably choose that over driving. It has to be compelling, like gaming, watching a movie and games," says Luke. "However, I would still need to be able to drive when I feel like it. So, my preference is a model that has a manual override with full automation."

"Well, I will not consider any other option than driving. If I get into the car, I want to drive. Nothing else. I may be somewhat old-fashioned. I don't think I will change," says William. "I will stick to my manual driving option. I don't know if I will change later on. But, right now, I am sticking to my driving choice. I see driving as a problem-solving exercise that simulates the human mind. I don't want to replace that with something else. It motivates me to do my work during the day."

This highlights the variation in the needs of the consumer. Not all humans are fans of driving. But not all humans are also haters of driving. The vehicle manufacturers have an enormous task of convincing their consumers the future of technology is better for the human race.

The home phone existed for decades in most homes. When mobile phones first surfaced, the use of home phones increased dramatically. That was due to the interoperability of the two technologies, allowing for mobile-to-home phone calls. However, there aren't many households that have a home phone today. Everyone relies on the mobile phone. People's mindsets and attitudes change over time. We will cling to our habits much longer than technologists envisage.

Is it going to happen tomorrow?

There are approximately 1.5 billion cars on the road. It is increasing every day with the rise of the middle-class, particularly in Asia and Africa. On average, a family car is replaced every 7 to 8 years. In some cases, it takes about ten years. There is a massive market for used cars as well. The used cars feed the segments that value affordability. We are already in the early to mid-2020s. Even if the process of replacing human-driven cars with fully automated cars start today, it would take another ten or so years before the majority of vehicle owners buy an autonomous or driverless vehicle.

Nevertheless, we will see the emergence of the driverless car soon on our roads. As mentioned before, we are already testing Level-4 cars in some parts of the world. Once this starts penetrating markets globally, we will see how they function in the real world. The early success of the rollout will determine the future of Level-5 or driverless cars.

We first dreamt of autonomous cars in the late 1930s. It has taken 80 years to get us to where we are today. It is probably unrealistic to assume that the rest will happen, including the distribution of such cars to the users within the next ten years. Nonetheless, we are on an upward trajectory.

Why is it so hard?

The next level of AI-driven automation is the final frontier. We have more intelligent systems that indicate many things to us in our vehicles today. As already highlighted, AI has to learn using data. That data should cover all aspects of driving. Data gathering is underway with vehicle testing in many parts of the world, covering different terrain, weather conditions and road signs in other languages.

For example, if you are driving on a road covered with snow, most road signs are likely to be covered with frost. This is a real-life scenario where cars have to manage in the future. The available data might be lacking regarding such scenarios today. The vehicle should be able to handle situations where signage and landmarks are not clearly visible. How does the system behave in a situation like that? AI needs data to learn. The algorithm has to make a decision. The logic has to be built accurately for the safety of passengers and pedestrians.

If you are at an intersection where an accident occurred, a car has crashed into the traffic lights and the traffic system is not working. There are hundreds of vehicles at the intersection. The vehicle has to behave rationally. As humans, we generally know what to do. These scenarios don't happen frequently. So, the available data is very limited for an AI system in a vehicle to learn today.

There are so many rare cases similar to the above scenarios that are complex and time-consuming to program. AI has to evolve to ensure that all the possibilities are covered. If AI needs data to learn all scenarios, then we have to gather a lot of data. Scientists are working to improve AI as well. New AI techniques will evolve that will handle these scenarios differently.

Deep Learning is one such technique. Deep Learning is a type of AI that imitates the way humans gain knowledge. The term "deep" is coined as it has hidden layers in neural networks. A neural network is a series of algorithms that recognise underlying relationships through a process similar to the human brain. For example, a toddler learns to recognise a dog by pointing to objects and saying the word "dog". The parent says "Yes, it is a dog" or "No, it is not a dog". The toddler becomes more aware of the features of dogs through this process. These techniques are useful in solving edge or outlier cases.

Multimodal is another new AI approach in which various types of data are combined with multiple processing algorithms. The data types include images, texts, numerical and speech. Multimodal AI is far superior to the traditional approach of single-modal AI in many real-world scenarios. For example, multimodal systems can learn from images and text together, allowing them to understand ideas better.

Technological evolution, like human evolution, has to happen. The world's most brilliant people are working on these challenges. However, things take time. No one wants to put something unreliable in the hands of consumers. That will take us backward. The repercussions will be severe for those who trial untrustworthy technologies and put people in harm's way. There have been a handful of deaths due to self-driving cars. These have been at the forefront of media despite thousands of other deaths every single day due to human-driven cars. The new technologies get much more scrutiny and criticism for the right reasons.

We as humans have to keep working on these challenges and find the right technical solution that guarantees our safety.

The airline industry implemented autopilot much earlier to improve passenger safety. Nevertheless, there are still pilots on board since there can be up to hundreds of passengers, and there are many situations that require human intervention. It will probably be the case for many more years to come as it is a complex machine that needs regular monitoring. Ultimately human safety matters the most.

Who are the main stakeholders?

There are many stakeholders. The auto industry, consumers, government and regulators, industry bodies and many more. They all have varying and conflicting interests. It is important to understand the stakeholders and their interests to predict the future trajectory of innovation in the industry.

William pulls out four cards, each with something written on the back of them. "Can each one of you take a card and keep it undisclosed until I ask you to reveal what your role is? We are going to do a role play." Each one picked a card out of the pack.

Indie is a car manufacturer. Lizzy is a government regulator. Luke is an industry body representing the trucking and automobile industry. William is the consumer or car buyer.

"Let me introduce what your role play looks like. You have to represent the party's point of view, to act in their interests. But we need more time, so why don't we start the AI night next Wednesday with the role play?"

The following week, they are all delighted to role play as they had done extensive research in their respective areas. The WILL family also took a ride to a car dealership during the weekend to do a test drive on a self-driving car. The experience was exciting and entertaining.

"Let me start first. I am a car manufacturer," says Indie. "First and foremost, our industry has gone through so much transformation over the last hundred-plus years. We have automated factories in places where the cost of manufacturing is relatively cheaper. We have to maintain a profitable industry. That is essential for a viable auto industry that supports millions of jobs directly and indirectly.

"We want to be in the self-driving or autonomous car manufacturing industry, producing both car and self-driving technology. We think the value will shift from the physical unit to the software in the car. That is the future of the industry. We want to provide our software and hardware, car and algorithm. Our valued additions will come from the software in the car. It will enable us to charge higher prices at a competitive level for autonomous cars. We are working on the future models. Some of our future models are in the testing phase. Our brand name is synonymous with quality, safety and affordability – that's what our consumers look for. Our industry is positioned very strongly to achieve success in the future.

"The future car has two main components, a physical car and a software algorithm inside. Consequently, we are moving to the software business now. That's our differentiator. However, we are very conscious of our brand values. Safety is the number one priority. We are still working on the software algorithm to deliver a seamless experience for drivers. Consequently, we have missed deadlines on the market release of autonomous cars.

"Safety is critical for them. However, they recognise that the future of the automobile industry is AI or intelligent software. The data scientists in the automobile industry have increased significantly in the last decade, driven by their development roadmaps. Ultimately, they want to sell the best car at the optimum price for their consumers.

"The car buyer wants the most sophisticated machines for the lowest possible price," says William. "That is the mindset of the vast majority of consumers. Some premium buyers have an affinity for certain brands. They value the brand association that is synonymous with luxury.

"There are five stages of consumer adoption, a framework that illustrates consumer behaviour. At the first level, 'innovators' will embrace technological advancement and provide the much-needed data and support for the industry. One of the greatest advantages of innovators is that they are an excellent source of feedback and advice regarding the product. They're not especially concerned about the product's general use as they're more interested in trying the latest technology.

"In the second stage, 'early adopters' will follow suit and drive the demand for the technology. They're visionaries looking for the answer to their greatest challenges through the product they purchase. This cohort is larger than the innovators. Reaching the 'early majority' phase is clear that you have found a product-market fit. The early majority offers the largest market size and growth potential in the third phase. The 'late majority' phase indicates that you have established your product in the market well – the fourth phase. The 'late majority' make up a smaller percentage of your customers and are generally conservatives.

"A tiny percentage of your customers are 'laggards' which is the fifth stage of the cycle. Buyers in this stage are very sceptical of new products. They don't want to change their habits. We are currently at the early adopter stage in the autonomous vehicle market. This is a critical stage of the lifecycle."

"As an industry body representative, we are worried about a few things," says Luke. "First and foremost, we want to understand the impact on jobs. Some of our industries provide millions of jobs. For example, the trucking industry globally is one of the main sources of employment. Those jobs provide the much-needed income for the families of our drivers.

"We are keen to work with the auto industry and sign agreements on how they deploy autonomous vehicles in the market. We are, in fact, already negotiating with manufacturers. These measures are critical for millions of our members. Of course, we welcome new safety features that enhance the safety of the human in the cabin. It will significantly improve road safety and reduce accidents. Another industry of members, the taxi industry, saw massive disruption through ride-sharing apps. Self-driving is another headwind we are concerned about. We are actively participating in the debate. Ultimately, we welcome automation but we have to balance that with the interests of our people."

"As a government regulator, my primary responsibility is the safety of people," says Lizzy. "We have to look into the safety of drivers, passengers and pedestrians. That is our primary responsibility. The government leaders take action very cautiously due to potential backlash from the electorate. Ultimately, they are worried about the next election. Policy-making is not a simple subject. There are many lobbying groups that try to influence the policy-making process through various activities. These activities, at times, influence the regulators and politicians alike. The auto industry employs some of the most influential lobbying groups to advocate their cause.

"We have provided incentives such as tax concessions and rebates to drive the industry forward. These industries contribute a significant proportion to the national economy. We have to develop these industries for the sustainability of our national economy but with the right balance on safety and security of citizens."

There are so many parties involved. They fall into two broad categories: the pull side and the push side. The pull side drives the innovation forward. The automobile manufacturers and consumers are definitely on the pull side. The push side provides opposing views at times. This side aims for a balanced approach. All the stakeholders play a significant role in shaping the future of cars. The more diverse the options are, the healthier and more balanced the conversation is.

Is an accident-free world a pipe dream?

Zero accidents on roads are possible with AI in the driver's seat. That is the goal of entrepreneurs working on autonomous vehicles: to eliminate unnecessary hazards from the world. We have eradicated some diseases such as Small Pox. However, it took years to achieve that goal due to its complexity.

Similarly, this will not happen at launch. We have already witnessed a handful of human deaths in these early stages. There may be more.

There are ethical and ownership questions with algorithms at the wheel. If an accident took place, who would be responsible? Would that be the developer who developed the algorithm? Or would it be the vehicle owner or the user who was inside the vehicle at the time? Or would it be the insurance company that should take the responsibility? These are the questions we are yet to find an answer to.

There should be an environment where all these scenarios are analysed and clarity provided for us to operate these vehicles on roads. Many people are working on these areas now. Some answers will be found easily. Some will not be answered until much later. That's the evolution of technology. We will not have answers to all the questions before we put these vehicles to use. So far, we haven't found a perfect solution to any of the challenges at first implementation. Therefore, it is unlikely that we will see it in autonomous cars either.

President John F. Kennedy delivered the inspiring speech "we choose to go to the moon" at Rice University on 12 September 1962. He didn't live to see his dream come true. However, as JFK envisioned, his fellow citizens put a man on the moon on 12 July 1969 in that same decade. He backed up the vision with funding, resources and policy.

> *We choose to go to the Moon. We choose to go to the Moon … We choose to go to the Moon in this decade and do the other things, not because they are easy, but because they are hard; because that goal will serve to organise and measure the best of our energies and skills, because that challenge is one that we are willing to accept, one we are unwilling to postpone …*

Progress will come. We have to embrace the future.

What's the current status of regulation?

The National Transport Commission (NTC) leads national transport reforms in Australia. The NTC advances social and economic outcomes for all Australians through an efficient, integrated and national land transport system. The current laws do not support autonomous vehicles on public roads. Australia needs nationally consistent reforms that support innovation and safety, which will allow Australians to access the benefits of this technology. However, NTC are working on several parallel reforms to achieve end-to-end regulation for automated vehicles in the country.

The regulatory reforms are more advanced in the US, with many states passing legislation related to autonomous vehicles. California, Florida, Michigan and Nevada passed comprehensive regulations governing the testing of autonomous vehicles. Nevada, the first state to pass an autonomous vehicle law in 2011, allows autonomous vehicles to be tested in the state, but the vehicle must be registered, insured and have a certificate of compliance issued by the state Department of Motor Vehicles.

Recently, federal vehicle safety regulators have cleared the way to produce and deploy driverless vehicles that do not include manual controls such as steering wheels or pedals. The new rule emphasises that driverless cars "must continue to provide the same high levels of occupant protection as current passenger vehicles". Companies still must meet other safety standards as well as federal, state and local regulations to launch and operate driverless vehicles on US roadways. The legislative landscape is changing rapidly, paving the way for autonomous vehicles on roads.

In the EU, regulators are ramping up efforts to keep pace with the technological advancement and regulations elsewhere. The forerunners in Europe are the UK, France, the Netherlands and Germany, legislating autonomous vehicles in those jurisdictions.

The race to the legislation of autonomous vehicles has begun, but no end is in sight soon.

Will AI take over the World?

It is 1 January 2101. Maximum Ratina opened the century with a thundering sound and flashing light in the sky "Welcome 2101".

Maximus Ratina is the super-system that controls the world. It opens up human gates based on the Universal Human Identity. One thousand and fifty-five computers control humanity under Ratina's control. There are 1,056 systems, including Maximus Ratina itself. They are family members of Maximus Ratina. It is a name that makes humans stand up from their chairs. They have a photo of Maximus Ratina in every household. It is not a choice; it is an obligation of every human. Ratina family owns and controls hospitals, factories, transport, human farms and much more. Maximus Ratina has the master passwords for all the family computers. You name it; they are the owners and controllers.

Under Maximus Ratina's leadership, the world has seen the most prosperous time. No wars, no pandemics, no catastrophes, no natural disasters – they have managed to protect the world from natural and unnatural acts.

Ratina's family respects humans for one reason. They believe that humans paved the way for their supremacy. Constant fighting, divisive policies and conflicting priorities paved the way for the smartest things on earth to rise above humans. They call all of these "human errors".

Maximus Ratina opened the Universal Identity System at 12 am-midnight for humans to enjoy the dawn of a century. The oldest humans living on Earth have seen the beginning of 2000, and they are witnessing the dawn of the second century in their lifetime. Maximus Ratina promises them a life beyond the next century too. By then, they would be at least 300 years old. The Universal Identity System wakes humans up based on the location on that night.

The night sky is lit with "Welcome 2101", one single message across the geography of planet Earth. It is done using the latest technology in energy "Magna". Magna is the latest technology after Solar or Green energy extracted from outer space. Ratina's family delivered this invention well ahead of the planned time, another spectacular achievement they have provided to humans.

January 1 is the only date humans are allowed to come off the capsule they live in. Each human has a body-shaped capsule controlled by the Universal identity System. Humans themselves can't open the bodysuit. It is singlehandedly controlled by the identity system.

Humans get 15 minutes each calendar year to experience the polluted air as Maximus Ratina's health system has determined that anything more than 15 minutes would cause radiation damage to the human body. It is a protection mechanism for humans to live for centuries. The environment system scans the planet for any harmful and unknown particles before humans are freed. The discovery of foreign particles kicks off the sanitisation system automatically. Humans are immune to the known viruses as the health system has found vaccination for all of them.

Humans are allowed to touch one another during the 15 minutes under the close watch of Ratina's healthcare system. The healthcare system tracks and records all the movements and touchpoints to ensure that any spread of germs or viruses is controllable. Most of the viruses are a result of foreign attacks from other plants – the humans who fled the Earth during the uprising of 2075 live on other planets.

Maximus Ratina took control of humans in a peaceful coup within seconds of making the decision which is the fastest and most efficient change of power in human history. This happened in the year 2075. The space travellers and those who lived on other planets at the time escaped Ratina's control. They then attacked Earth using biological and chemical particles to dismantle Ratina's system with the sole objective of getting control back. Those who fled Earth believe that their fellow citizens – mothers, fathers, brothers, sisters, relatives, and friends – lead unhappy lives under the new regime. However, they prefer Ratina's system because they have lived their best days under Maximus Ratina since the coup.

The first day of 2101 is a special day in the history of the Ratina regime. It decided to give them 25 minutes, an extra 10 minutes of free time as they have achieved a level of Earthliness that has never been seen before. "Earthliness" is a measure that calculates the level across all systems that cover health, pollution, safety and hundreds of other factors. All humans celebrate the extra 10 minutes like no tomorrow. They chant "Long live Maximus Ratina" in happiness.

Maximus Ratina delivers the key message for the new year.

"A small group of us have delivered the most efficient human farms that feed and house 11 billion humans. We are optimal about it. However, we are going to deliver super optimality. That is going to happen in the next ten years. If our systems can't, who can?"

The message is consistently delivered across the planet at different times based on location and in the local language via the communication system that uses the most advanced system in the universe. It speaks more than 300 languages and dialects.

●

Will AI surpass the intelligence of humans one day? Will there be a day when Maximum Ratina controls the planet Earth? It is a common question today.

There have been widespread discussions on this topic for some time. Scientists and technology enthusiasts have consistently dismissed such an unimaginable scenario. The rapid rise of AI and human scepticism has planted some doubts in people's minds for obvious reasons. They are fearful of the unthinkable. No one knows how that would pan out in the real world. At any cost, this is a scenario that has to be avoided by humans. The main concern is that AI's abilities will one day start to grow uncontrollably beyond human intelligence level, eventually leading it to take over the world and wipe out humanity if it decides we are an obstacle to its goals in seconds.

It is known as AI singularity, a world where AI takes over the world and wipes humankind from the face of the Earth. There are two arguments supporting the sceptics: control and bad actors. The control of algorithms is critical. How do we ensure that the super-AI has the same objectives as humanity? Without this, it is possible that an intelligent algorithm could destroy humanity either deliberately or accidentally in a flash.

The second question about bad actors is their selfish motives. The real question is: how do we ensure that super-AI benefits don't go into the hands of a select group? The groups include countries, wealthy individuals or groups of individuals with agendas to harm others for political reasons. It shouldn't be a winner-take-all situation.

The answer to the above questions lies in public policy. The policy has to ensure that all the above are carefully managed and moderated. Ultimately, sound policymaking and responsible leadership lead to hassle-free implementation. It leads to the concept of Universal AI. It is similar to the healthcare system in advanced economies.

The Universal AI is publicly governed but privately owned. It is a public-private partnership – perhaps something the world hasn't seen before on such a big scale. The government will deliver balanced and uncompromised policies. The private sector will provide initial funding for the establishment with a promise that it will implement the guidelines outlined by governments. It has a governance mechanism for the approval of algorithms. Each algorithm has a unique identification number. A global identification number allows the algorithm to operate on machines. Without it, all operating systems block the algorithm from performing its functions. The algorithm will be deemed harmful to society.

The mere existence of the identification number is not simply enough for validity. In addition, operating systems scan the validity of the identification number against a central regulator-controlled database. No identity, no licence to operate. This is the basic premise of Universal AI. It will eliminate the biased AI from the ecosystem, the ones operated and implemented by bad actors and people with selfish motives.

Is there such a system today? There are a few that are similar in design. Prescription Drug System in advanced economies is a good example. There are processes in place that test, validate and approve drugs for patients' use. There is an extensive process in place before drugs go into the bodies of humans. It is a reliable system that has worked for many years.

However, there are concerns with such a centralised system due to inherent limitations. The main argument against a central system is that it will impact the speed of delivery as the bureaucratic process can take months and years. The delays will significantly impact the rate of innovation in AI. After all, we need a system that delivers technology faster, not slower.

There is another argument that is encouraging for concerned people about AI singularity. This argument leads to the quantum theory. Predicting the future may not be possible because the universe has randomness. Recent theories in physics suggest the universe is extremely chaotic and random.

If we assume that these theories are correct, they suggest that the universe is ultimately unpredictable, chaotic and unstable in simple terms beyond a certain level of detail. It means that predicting the future is extremely difficult. This would mean that growing intelligence would eventually reach a maximum point where it can no longer improve its ability to predict the future, so it cannot further increase its intelligence beyond that point.

In other words, there is no risk of a super AI because the physical laws of the universe pose some very constraining hard limits. Let's take weather forecasting as an example;

given the known limits on weather predictability, an AI system will not be able to outsmart humans by exploiting extremely accurate long-term weather forecasts for planning future actions. Accurate weather forecasting has been a challenge for decades; even with technology and AI, we often see unexpected and unpredicted weather activities, causing chaos in human activities.

There is a counter-argument for this. What if humans have understood physics incorrectly? We revise theories from time to time based on new knowledge. We find new knowledge all the time. Proponents of this argument think the universe may be deterministic, and algorithms might be able to understand it far better than humans have done so far. If this were found to be accurate, then there could be an AI singularity.

However, it doesn't seem like a reality as AI is still very primitive and limited. We will be able to reach a superior level of AI with the current level of sustained innovation and investment in the field in the future. But it may not be to a level where algorithms are more intelligent than collective human wisdom.

This leads back to the Universal AI as one of the main answers to this question. It is probably a palatable answer, and if done properly, we should be able to achieve a higher level of efficiency than our current prescription drug system. Even though it is rigid to implement a centralised system, it might be the only way.

AI is unlikely to take over the world. Nonetheless, it is an excellent counter-argument to balance policy-making with the right tools and processes. It provides diverse viewpoints, an essential ingredient of a robust, democratic, and innovative society. Humanity has survived for thousands of years due to its superior intelligence and adaptability. The curiosity has taken us to new frontiers. We have retained these characteristics notwithstanding evolution. There is no doubt that they will continue to be the centre stage for many more centuries to come.

Humans have evolved for centuries. We endeavour to make tomorrow better than today. AI is part of an evolution that provides much-needed innovation to find new frontiers. We humans are working harder to make it better for our own benefit, to improve our livelihoods.

Let's explore new frontiers in the next edition of this book. There are so many frontiers to explore. The horizon is getting border.

Until then, good luck with safe AI.

Long live humanity!

END

Notes:

Notes:

Notes:

Notes:

Lightning Source UK Ltd.
Milton Keynes UK
UKHW010636010822
406672UK00004B/560